Lawyers' Skills

Lawyers' Skills

Julian Webb

Professor of Legal Education, University of Warwick

Caroline Maughan

Principal Lecturer in Law, University of the West of England, Bristol

Mike Maughan

Senior Lecturer in Organisational Behaviour, University of Gloucestershire Business School

Marcus Keppel-Palmer

Senior Lecturer in Law, University of the West of England, Bristol

Andy Boon

Dean of the School of Law, University of Westminster

OXFORD
UNIVERSITY PRESS

OXFORD
UNIVERSITY PRESS

Great Clarendon Street, Oxford OX2 6DP

Oxford University Press is a department of the University of Oxford.
It furthers the University's objective of excellence in research, scholarship,
and education by publishing worldwide in

Oxford New York

Auckland Cape Town Dar es Salaam Hong Kong Karachi
Kuala Lumpur Madrid Melbourne Mexico City Nairobi
New Delhi Shanghai Taipei Toronto

With offices in

Argentina Austria Brazil Chile Czech Republic France Greece
Guatemala Hungary Italy Japan Poland Portugal Singapore
South Korea Switzerland Thailand Turkey Ukraine Vietnam

Published in the United States
by Oxford University Press Inc., New York

First published 1993
Reprinted 1993
Second edition 1994
Reprinted 1994
Third edition 1995
Fourth edition 1996
Fifth edition 1997
Sixth edition 1998
Seventh edition 1999
Eighth edition 2000
Reprinted 2001
Ninth edition 2002
Tenth edition 2003
Eleventh edition 2004
Twelfth edition 2005
Thirteenth edition 2006
Fourteenth edition 2007

British Library Cataloguing in Publication Data
Data available

Typeset by Laserwords Private Limited, Chennai, India

Printed and bound in Great Britain by William Clowes Ltd, Beccles, Suffolk

ISBN 978–0–19–921218–7

10 9 8 7 6 5 4 3 2 1

OUTLINE CONTENTS

DETAILED CONTENTS

ONLINE RESOURCES TO ACCOMPANY THIS BOOK...

Online Resource Centres are developed to provide students and lecturers with ready-to-use teaching and learning resources. They are free-of-charge, designed to complement the textbook and offer additional materials that are suited to electronic delivery. The Online Resource Centre to accompany this book can be found at:

www.oxfordtextbooks.co.uk/orc/lpcskills07_08/

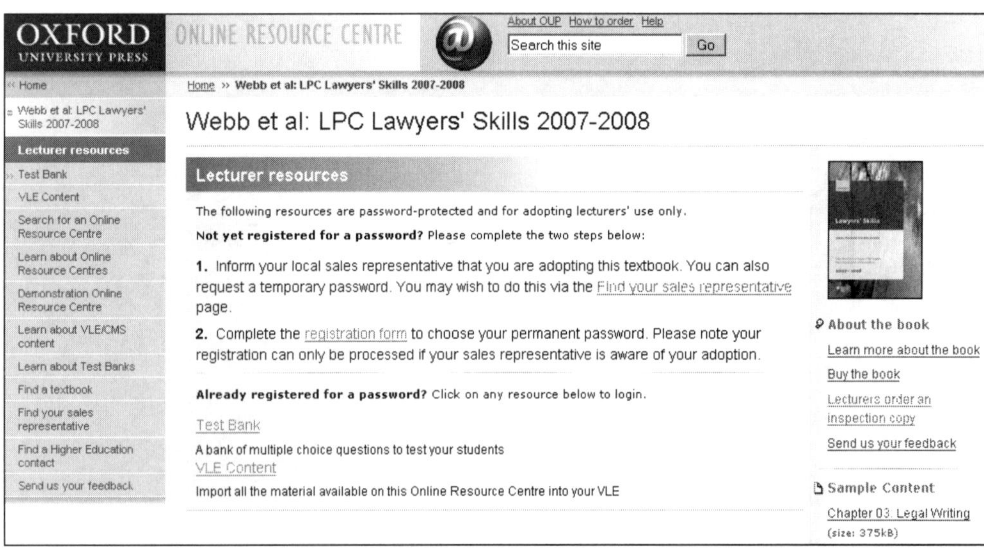

Lecturer resources

Password-protected to ensure only lecturers adopting this book can access these resources, each registration is personally checked to ensure the security of the site. Use these resources to complement your own teaching notes and the resources you provide for your LPC students.

Registering is easy: click on 'Lecturer Resources' complete a simple registration form which allows you to choose your own user name and password, and access will be granted within 48 hours (subject to verification).

Test bank of multiple choice questions

Using your lecturer password, you can gain access to a fully customisable bank of multiple choice questions offering a versatile way to test your students' knowledge and understanding of the skills covered in this book. The questions are downloadable into Questionmark Perception, Blackboard, WebCT and most other virtual learning environments capable of importing QTI XML. The questions are also available in print format.

OXFORD
UNIVERSITY PRESS

LPC Lawyers' Skills 2007-2008 Test Bank - Chapter 02

1 of 9

Which of the following is **not** one of the main purposes of a client interview?
- Create a feeling of confidence in your client as to your competence and commitment to their case.
- Gather information in order to identify ways in which the client's aims can be achieved.
- Help your client to identify precisely what they want from the situation.
- It is a means of getting all the information you need in order to decide what you are going to do about the situation.

2 of 9

Which of the following is a correct description of the WASP approach?
- The WASP approach is to sting your client for as much money as possible.
- The WASP approach is to get to the heart of the issues as quickly as possible.
- The WASP approach is to structure a client interview appropriately.
- The WASP approach is to remind you to begin and end the interview properly.

3 of 9

In an interview which one of the following describes the best use of closed questions?
- Closed questions are used to ask for specific, unambiguous information.
- Closed questions are used to check whether the interviewee is telling the truth.
- Closed questions are used to stay in control of the interview.
- Closed questions are used to indicate that it is time to move on to another part of the interview.

ACKNOWLEDGEMENTS

Extracts from the Legal Practice Course Written Standards are reproduced with kind permission of The Law Society.

1 Introduction

1.1 What this guide is about

This book is concerned with the lawyers' skills that underpin practice as a solicitor and which are introduced during the Legal Practice Course (LPC). It is structured to take you through those skills in more or less the same way that you will encounter and use them in practice situations.

It therefore begins by introducing you to client interviewing. Interviewing and advising are obviously fundamental skills for the lawyer, and also help us to focus on aspects of oral communication that are relevant to other professional tasks, such as negotiating and advocacy. Because they are quite complex, composite activities,these other oral skills are not introduced until **Chapters 7** and **8**, chiefly as a foundation for the more specialist training that you may undertake during your training contract and beyond. This first chapter provides a structured introduction to the skills and processes involved.

The focus then shifts to writing and drafting, which are probably the skills most used by trainee solicitors. In **Chapters 3** and **4** we examine how you can use language effectively in a variety of contexts: writing letters, drafting a contract or will, or drafting documents in litigation.

We then turn to the foundation skills of legal research and problem solving. Legal research is another activity that will probably occupy a significant amount of your time as a trainee. In **Chapter 5** we look at how practical legal research both builds on, and yet differs from, the kinds of research activities you should have undertaken at law school, and we provide you with some tools and strategies for undertaking practical legal research. **Chapter 6** on problem-solving similarly focuses on how practical legal problems differ from the 'textbook' problems of the academic stage, and offers a problem-solving model and techniques that will support both contentious and non-contentious practice.

Finally, we have chapters which are primarily concerned with enhancing your ability to manage and develop your professional life. The first, on workload management, is intended to help you adjust to the demands of a professional career in terms of task and project management skills, and also by encouraging you to think about the important issue of stress management. The closing chapter, on continuing learning, recognizes that in an increasingly complex and changing world, the public and the profession are entitled to expect solicitors to show a personal commitment to continuing learning and professional development. In this context, the LPC is only the first step in your professional education, and we attempt to show here how you might begin to build on that foundation.

In the remainder of this Introduction we want to make a few points which will help you make the most of the skills training you receive on the LPC.

1.2 The nature of legal skills training

This Guide is a supplement to rather than a substitute for the skills training you will undertake on the LPC. If you have not previously experienced skills based learning, you need to be aware that it creates a rather different learning experience from the more conventional teaching approaches with which you may be more familiar. Skills based learning is characterised by the following features:

(a) It is participative. You cannot learn skills by reading books, and not wholly by observing others. You learn skills by doing.

(b) But doing is not enough! If you are to learn from doing, you must be prepared to analyse and reflect on your experiences: how it felt to be engaged in a particular task; how others reacted to you; how you and others thought that you might improve your performance of that skill. This is why you will be asked to get involved (in various capacities) in exercises and simulations, and encouraged to review your own performance, as well as to receive feedback from peers and trainers.

(c) It is important to undertake further opportunities to practice your skills. Time on the LPC is limited. You will get some opportunities to use your skills in different settings, but you will benefit enormously if you can manage to practise and analyse your skills outside of formal workshops and classes, and take time out to observe experienced practitioners in action.

(d) At the same time 'skills' does not just involve learning from experience. There is a growing body of concepts, theories and techniques which can help you develop competence and aid in analysis and reflection. One function of a guide such as this is to introduce you to the theory underpinning the skills.

In summary then, the aim is for your learning to become deeper and more personalised as you work through a continuing process of experience, reflection, analysis and revision of your skills.

1.3 How to use this guide

If follows from what we have said about the nature of skills-based learning that we expect you to use this Guide more as a workbook than a conventional textbook. It is not something you sit back and read passively! To help you engage with the materials we have included two particular features in these materials.

First, you will see scattered through all the chapters a number of text boxes containing instructions. These boxes contain short tasks which are intended to make you stop and think about the issues/skills/processes being discussed. Please do not skip over them, you will get more out of your reading if you work through the activities they contain.

Secondly, where possible (and appropriate) we have included some more substantial exercises. Some of these may be incorporated into your workshop activities (for example, the contract exercise at the end of **Chapter 2**, 'Interviewing'), or they may be left as an additional resource for you to use to develop your skills in your own time.

While most of these tasks and exercises can be completed individually, some require, and others would certainly benefit from, discussion or collaboration with others (particularly if there are opportunities for role-playing). In some instances, we have indicated where collaborative work would be an advantage, but it is probably fair to say that we all tend to the view that collaborative learning is generally to be encouraged, as it almost invariably creates opportunities to share useful ideas and experiences.

The materials in this Guide also contain two other features with which you may not be familiar.

Every chapter opens with a statement of learning outcomes. These are there for two purposes:

(a) to help you identify at the outset what each chapter is trying to achieve; and

(b) to enable you to check that your learning is on target to meet those outcomes.

Where relevant, we have also included the Law Society's Written Standards for the LPC. These indicate the areas that are to be assessed and provide more detailed indications of the skills and attributes you will be required to achieve competence in. You will see that there are written standards for all the major skills on the LPC, i.e., advocacy, interviewing, writing and drafting, legal research and (to some extent) problem-solving. There are a number of skills areas therefore (notably negotiation, work management and continuing learning) which are, at present, non-assessed. We have introduced some of these skills into the Guide as part of a package of changes which we think will help you to develop a wider vision of preparation for the realities of modern practice. We hope you find them a useful addition.

2

Interviewing and advising

2.1 Learning outcomes

This chapter deals with the skills of interviewing and advising clients. In this chapter we focus particularly on:

- the purpose of the initial client interview;
- the structure of an effective client interview;
- preparation for the interview;
- appropriate client care;
- listening and questioning techniques;
- providing appropriate advice and information;
- establishing a professional relationship with the client.

2.2 What are interviews for?

The purpose of interviews is to get and give information and decide what you are going to do with the information you have got. It is also an opportunity to establish and maintain high standards of client care. We cannot overstate the importance of first impressions that clients get from face to face meetings.

Although we are dealing principally with interviewing and advising clients, remember that you will also have face to face meetings with other lawyers, other professionals, witnesses and other parties, where your interviewing skills will play an important part. Whoever you are meeting, all your interviews will have a similar underlying structure.

When you interview clients you will usually be aiming to:

(a) help your client identify precisely what they want from the situation;

(b) gather information in order to identify ways in which the client's aims can be achieved;

(c) help your client to reach decisions about the most appropriate way to get what they want; and

(d) create a feeling of confidence in your client as to your competence and commitment to their case.

2.3 How should you prepare for an initial client interview?

> Write down the main ways in which you think an interview with a client would be different from an interview with another professional (e.g., a social worker, a police officer, a DTI official, a medical specialist).

We think some of the main differences would be:

(a) other professionals will have a detached and analytical view of the situation and will not be so emotionally involved as a client;

(b) they are not running any financial risk in participating;

(c) you might expect a high degree of shared knowledge about the way cases proceed, standards of proof, timescales, cost etc.; and

(d) other professionals will be more likely to understand legal terminology in their own area of expertise.

You may well have thought of other differences. The implications of understanding these differences are important for the way you prepare for an initial client interview. You cannot make any assumptions about the client's knowledge, about the emotional impact that pursuing a case might have on the client, about the nature or level of financial or other risk the client is willing to accept. You don't even know whether the problem the client is coming to you with is capable of a legal resolution. Often all you have is a general indication of the client's concerns, from some initial contact between your organisation and the client. How the matter proceeds depends on your skills as an interviewer.

2.3.1 Preparing the environment

> If you were a client who wanted to explore possible solutions to a problem you had, what minimum expectations would you have about where the interview took place?

As an absolute minimum, we would want the following:

(a) to speak in private without our conversation being overheard by others in the room, passers by, people in an outer office, etc.;

(b) not to be interrupted by other people, telephones ringing etc.;

(c) to have enough time to discuss the matter without others wanting that room, for example;

(d) to have the solicitor's full attention;

(e) to sit in reasonable comfort;

(f) no physical barriers between us and our solicitor to impede communication;

(g) not to be kept waiting.

The bare minimum, then, would appear to be a comfortable, quiet room where you won't be disturbed for the duration of the interview and a room that enables you to be as formal or informal as you and the client feel comfortable with.

2.3.2 Preparing yourself

> You have an appointment with a client, Mrs Tyler, who is coming to see you about an accident she has had at work. How would you prepare yourself for the interview?

A word of warning! Don't assume that this really is the client's problem. You won't know until you have carried out the interview and have had the opportunity of exploring her concerns in detail. The information you have is merely a starting point. You may therefore decide that there is not a lot of point in researching the law in any great depth. So what should you do?

One thing you might begin with is to find out if this client has used your firm before. If so, you may be able to get some information about her and about other issues the firm has dealt with. You might also speak to any of your colleagues who have met this client previously. The more information you can garner about your client, the better prepared you will feel. However, in many cases you will know virtually nothing.

What you can plan is the interview structure, so that you can come at the main issues quickly, professionally and comprehensively. For example, there are some common features to any client interview. The client must have the opportunity to say what he or she needs to say. You need to provide advice on the legal issues. You also need to alert the client to the financial implications of taking on the case and confirm that the client wants to instruct you.

In order to ensure that nothing of importance is left out, we suggest you use the WASP approach to planning, structuring and carrying out your interview. WASP is an acronym which breaks the interview into four parts. It stands for:

(a) Welcome;
(b) Acquire information;
(c) Supply information and advise;
(d) Part.

2.4 How does the WASP approach work?

2.4.1 Welcome

It is critically important to get this part of the interview right. It may very well be the first meeting the client has ever had with a member of your firm. It may even be the very first meeting with a solicitor. Make sure that when your client arrives they are greeted and made comfortable and that you do not keep them waiting. In our experience, many students refer to this process as the 'meet, greet and seat' part of the interview.

At this stage it is useful to let your client know what to expect from the meeting. You may wish to tell them:

(a) the purpose of the meeting—that is, to get details of the situation from your client, give legal advice, discuss options and provide information on costs etc.;
(b) information on the service levels your firm provides, such as how frequently the client can expect to be updated on the progress of their case. We discuss this in more detail in the 'S' part of WASP.

EXERCISE 2.1

Work with a partner. One of you takes on the role of the solicitor, with the other role playing Mrs Tyler.

(a) Practise the welcoming phase of the interview up to the point where the solicitor begins to ask about the client's problem. Remember to arrange an appropriate environment as far as you can.

(b) Discuss what went well and why and what worked less well, and why.

(c) Switch roles and carry out the welcome phase again.

(d) Based on your reflections and feedback, write a brief set of guidelines for opening an interview.

(e) Compare your guidelines with those of other pairs. What do you find?

Typically in this kind of exercise, participants discover the importance of things like:

(a) the impact of non-verbal communication on the interaction. This includes things like whether or not someone smiled; the degree of eye contact; physical proximity and touch;

(b) appearing confident;

(c) appearing sympathetic and friendly;

(d) speaking naturally and sincerely (hard to plan for, but it comes with experience);

(e) not hiding behind a desk; and

(f) not making your introductory remarks too long, so that several minutes elapse before your client has the chance to say anything.

The 'Welcome' part of the interview is very important. It gives you the opportunity to make a good first impression on your client. It affords you the opportunity to let your client know what to expect from the interview and to establish an appropriate level of formality for the proceedings. Before moving on to the 'Acquire' part of the interview, your client should be comfortable and ready to discuss their concerns with you.

2.4.2 Acquire information

You do this by inviting your client to tell you why they have come to see you and by listening to what they say.

2.4.2.1 Letting the client talk: questioning techniques

Ideally, you want your client to tell you everything in their own words. Some clients are perfectly capable of doing this with little prompting. Others need to be encouraged. Developing your questioning technique gives you the best chance of getting at all the relevant information. Most interviews fall on to a continuum. At one end we have what can only be described as an interrogation; at the other a free flowing two-way conversation. The closer to a conversation your interview is, the more effective it is going to be. In an interrogation, one of the parties is an unwilling participant. In a free flowing conversation, both parties are willing to communicate and do so openly. In the former, the agenda is totally controlled by the interrogator, whilst in the latter, the topics discussed are often wide ranging.

The danger is that if a client is unforthcoming, it is tempting to close questions down, or use leading questions, so that the interview tends more towards the interrogation than the conversation. Similarly, an inexperienced interviewer may move towards interrogation in order not to lose 'control' of the interview process.

How might this be avoided? The main focus of an interview is to get your client talking, explaining things in their own words and expressing their feelings. There are a number of techniques of questioning to help us in this.

To encourage your client to speak:

(a) Use open questions. These are the 'what, why, how, when, where' questions. Such questions are impossible to answer in a single word or with a shrug. The respondent has to frame the answer in their own words. For example, 'What happened? Why did you think that?' etc.

(b) Invite your client to talk. For example, 'Tell me about...', 'I'd like to hear a little more about...', 'Please go on', etc.

(c) Use sympathetic body language such as a smile and a nod to encourage your client to go on speaking.

(d) Summarise periodically to check your understanding and encourage your client to correct any misunderstandings; 'So the situation so far is that... Have I got that right?'

(e) Don't be afraid of silence. Sometimes interviewees need time to think through an answer, or to find the right words to explain themselves. The effective interviewer gives them that time. The ineffective interviewer jumps in and fills the silence with another question to encourage the client to say something. Often, this new question is a closed question which attempts to make the client's answer as easy as possible, but which restricts the available answers and reduces the opportunity for your client to use their own words. Like any conversation an interview is based on rules about taking turns. A silence on the part of one person may understandably, but wrongly, be interpreted as an indication that they have finished their turn or want to miss it. The temptation is to help them out and allow them to miss it. Often, silence means the client is struggling to find the right way to say something (e.g., something embarrassing or something which doesn't show them in a good light or something which is painful for them to talk about). Don't therefore think that silence is a vacuum that you have to fill. Your client will almost always find the words to say what they want to say.

Things that will discourage your client from talking are:

(a) Using closed questions. Closed questions are questions which require only 'yes' or 'no' or 'don't know' answers, or require very specific information, such as: 'You say you suffered an accident at work?'; 'Was this recently?' Too many questions like these will only move the interview forward uncertainly and may positively discourage your client from putting things in their own words. Not only that, but the initiative passes to the questioner who is tempted to frame the situation in his or her own terms and construct an interpretation which differs from that of the client.

(b) Using multiple choice questions. These are like closed questions in that they allow a very restricted range of possible responses. These are questions like: 'Did you report it to the supervisor or the manager?'; 'Are you after compensation or your job back or both?' The objections to this are the same as for closed questions.

(c) Using leading questions. These questions expect a particular answer, e.g., 'I don't expect you'll want to go back to that job, will you?' The problem with this type of question is that it suggests to the client that you have formed an interpretation of events which may not be the same as theirs, but which you are inviting them to agree with. They may well feel that the situation is being taken from their control.

Try to avoid leading questions. Closed and multiple choice questions have their uses though. For example, if you want to confirm your understanding of an event or an issue, they are perfectly permissible. Or, when you have heard the main thrust of the client's story, you can use a series of closed questions to probe the details. 'You've told me that you reported the accident. Was that to your manager or to someone else?'

EXERCISE 2.2

(a) Read the dialogue below.

(b) Identify the question types used by the interviewer.

(c) Say how each question affected the interaction.

(d) Could the interviewer have made the interview more effective? If so, how?

Interviewer:	Please come in and take a seat, Mrs Tyler.
Client:	Thank you.
I:	How can I help?
C:	Well, I don't really know where to start.
I:	I gather you had an accident at work. Is that it?
C:	Yes, that's right.
I:	What sort of work do you do?
C:	Well, I've hardly worked at all since the accident.
I:	I see. But what exactly was your job?
C:	I was a clerical officer for Social Services.
I:	Had you been there long?
C:	Eight years?
I:	Right. Tell me about the accident.
C:	It's hard to know where to start really. I . . .
I:	When exactly did it happen?
C:	Last December, about a fortnight before Christmas.
I:	Right. Let's get the details, then. Tell me exactly what happened.
C:	We were having an office move and my supervisor told me to help her move the photocopier from the table it was on, to just outside the lift, so the removal men could collect it. We had to manoeuvre it through the door . . .
I:	It's not normally part of your job to move equipment, then?
C:	Well, no, not really. But everyone was chipping in that day.
I:	Not everyone had an accident, though, did they?

You probably agree with us that this is not a particularly good way to get the information you need because:

(a) The lawyer doesn't allow the client to tell her story but, because she is a little hesitant, he tends to jump in with closed questions and diverts the client from the story she wants to tell. He might have been better advised to remain silent on some occasions to allow his client to gather her thoughts and frame her own answer.

(b) In his eagerness to get at some of the detail of the case, the interviewer doesn't seem to acknowledge his client's likely feelings in this situation. For example, he doesn't enquire about her injuries, or whether she is still in pain. He gives the impression of wanting to follow his (legal) agenda and may well forfeit the client's confidence as a result.

(c) The lawyer didn't seem to have thought about how best to get the information from the client.

2.4.2.2 Funnelling

He might have used a questioning technique known as 'funnelling'. This technique is used to develop the discussion in greater detail. If the lawyer in the exercise had been listening carefully to Mrs Tyler he would perhaps have picked up several cues, which he could have followed up. Funnelling would have helped him follow up.

The funnelling process works like this:

(a) Ask a general, open, preferably factual question (what happened?).

(b) Follow this up by asking about the context of the event (who, when, where, what else?).

(c) Then you may find it useful to ask about the dynamics of the event (how, why questions).

(d) Finally you can confirm your understanding by using closed questions.

The idea behind funnelling is to work from the general to the detailed and it can be shown as follows:

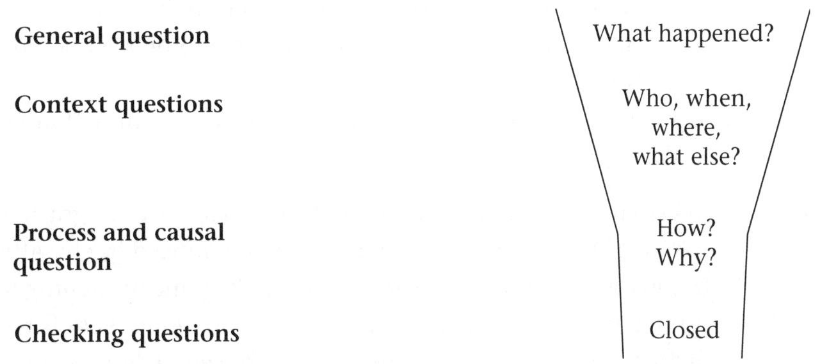

In some interviewing situations you will find it helpful to use the 'inverted funnel'. In the inverted funnel the sequencing of questions begins with very closed questions and gradually opens out to embrace wider issues. This may be useful in advising a client to decide on action to be taken. You begin by exploring the key issues first and build up the broad picture from the mass of detail. Once you have a clear picture of the facts—the estate of a client in a will case, for example—it may then be appropriate to consider the options that the client has in mind.

Let us see how the solicitor in the exercise might have funnelled had he been more receptive to what his client was saying. In a case like this, of course, the inverted funnel approach is not appropriate. We have no foreknowledge of the details of the case and need to get a general picture of what happened before we can focus on relevant issues.

We can start at the point where the solicitor asks Mrs Tyler to describe the accident:

I: Right. Tell me about the accident

C: It's hard to know where to start really. I... (pause) We were having an office move and my supervisor told me to help her move the photocopier from the table it was on, to just outside the lift so the removal men could collect it. We had to manoeuvre it through the door... (pause). It was very heavy and we had to twist ourselves to get it through the door. We eventually got it through the door and outside the lift. I felt a bit of a twinge in my back then, but thought it was just having to carry something.

I: Can you describe exactly how you carried it through the door?

C: Well... (pause) We had to unplug the machine and one of us took each end. Then we had to turn it because it would only go through the door one way. I took one end and Fran the other but I had to go backwards through the door... (pause) I took most of the weight because I'm bigger than Fran. Then, because it wouldn't go through the door with me holding it, I had to put one hand on top and the other underneath. That meant me having to twist a bit to get it through and look where I was going.

I: What was Fran doing at this point?

C: Well she was on the other end. She didn't have to twist or anything because she was walking forwards.

I: Why is it that you and Fran did this when there were removal men around?

C: The office was a bit cluttered because we were all turning out our stuff, so Fran said we should move the photocopier while we still had a gangway to the lift.

I: You said Fran is your supervisor?

C: Well, more of an office manager, really, but she used to be my supervisor. We've worked together for a long time.

I: Were you normally expected to lift equipment?

C: Well, we usually move small things if we have to.

I: Have you ever been asked to move a photocopier before?

C: No.

I: And have you ever been asked to move anything that heavy before?

C: No, never.

There are clearly many more points an interviewer might have picked up, such as the nature of Mrs Tyler's injury, when she first noticed it, etc. Alternatively, he might have begun another line of questioning using the same funnelling technique. The critical factor in funnelling is the need to listen to what your client is telling you. Only when you do this can you ask further questions effectively. This may sound obvious, but interviews are often less effective than they might be because the interviewer is a poor listener and doesn't pick up the issues to follow up.

2.4.2.3 Active listening

Listening is an active process and it is about analysing the information you receive and fitting that information onto the framework of your understanding. As someone speaks, you are constantly checking and modifying your understanding of the situation. As a lawyer you have developed certain cognitive frameworks which enable you to understand, for example, the law of negligence or employment.

Active listening is therefore an important part of the process of building an initial perception of the client's legal position. For example, when Mrs Tyler first tells you she has had an accident at work, the legal framework of negligence will spring to mind, and you will be listening for clues in her story of evidence of breach of duty of care, causation, etc. Your questions will be directed towards verifying that you are using the right framework and modifying that framework so that it applies specifically to this case. For example, part of your framework on the law of negligence will tell you that you have to prove causation. This would lead you to ask Mrs Tyler whether she previously suffered any back problems, or whether anything has happened to her since the accident which might also account for her condition.

Bolton in *People Skills* (Brookvale, New South Wales: Simon and Schuster, 1987, p. 30) discovered that 'even at the purely informational level, researchers claim that 75% of oral communication is ignored, misunderstood or quickly forgotten. Rarer still is the ability to listen for the deepest meanings in what people say.'

So how can you improve your active listening skills?

We can consider active listening as having three components:

(a) checking what the client is saying against your frameworks of understanding;

(b) following up points that you hear by appropriate questioning; and

(c) checking that you have understood by summarising and reflecting back what you have heard.

EXERCISE 2.3

Do this exercise with a partner.

(a) Both spend a few minutes preparing to talk about a controversial issue that you have strong feelings about.

(b) Give your talk to your partner.

(c) Your partner then restates what you have said, reflecting and summarising as accurately as possible. Your partner should try to reflect, not only the information in the talk, but their perception of your feelings about it.

(d) Then make any necessary corrections to ensure you have been accurately understood.

(e) Now swap roles and repeat the exercise.

(f) Finally, feed back to each other (and the workshop) what you have learned about active listening.

So far we have concentrated on getting information from the client which will help us understand the factual and legal aspects of their case. However, although this is important for the client, there is often also the need to have their feelings of anger, distress, disappointment, outrage, anxiety, etc., acknowledged. The effective interviewer does this through the process of active listening, by paying attention not only to the factual information that is being transmitted but also to the complex of emotions that lies behind what is said. Often when a client persists in emphasising something which to the lawyer appears irrelevant, it is because this represents or touches on strong feelings about the issue.

> You perceive that a client you are interviewing is very distraught, and that you are finding it difficult to get the information you need in order to advise them. How do you deal with this?

The main point to remember is that when we feel strong emotions, we need to express them. Until they have been brought out, very little else can be accomplished. Clients may be looking for you to acknowledge the strength of these feelings. It is important that you do this if you want to make progress with the client. One way of dealing with this is to confront this openly with your client rather than ignore it or pretend it is not happening. Look at the next exercise.

EXERCISE 2.4

Read the three short dialogues which follow and:

(a) say whether the interviewer responds satisfactorily and, if not,

(b) how the responses could have been improved?

Dialogue (A)

Mrs Tyler: Apart from the first couple of weeks after the accident, none of my friends at work have kept in touch with me and I don't know why.

Interviewer: You sound upset about that.

Mrs Tyler: Well, I am a bit. We were all very friendly, not just the staff, but the managers as well. They just seem to ignore me now. Even when I see some of them in town, they pretend they haven't seen me. One of them, Jamie, crossed the road last weekend so he wouldn't have to talk to me. It's got to the point now where I don't even feel that I can phone them.

Interviewer: Have you tried phoning any of them recently? Is there anyone in particular you could phone? At least then you could find out if they really are ignoring you.

Mrs Tyler: I could try, I suppose.

Dialogue (B)

Mrs Tyler: Apart from the first couple of weeks after the accident, none of my friends at work have kept in touch with me and I don't know why.

Interviewer: It's probably nothing. Does your employer know you've contacted a solicitor? If they do, your friends may have been told not to talk to you.

Mrs Tyler: (Upset) But they're supposed to be my friends.

Interviewer: (Gets up) Look, let me get you a cup of tea. Then we can get on.

Dialogue (C)

Mrs Tyler: Apart from the first couple of weeks after the accident, none of my friends at work have kept in touch with me and I don't know why.

Interviewer: Yes, that's very upsetting. Now about the accident. When exactly did you report it to your employer?

Opinions about the relative merits of the approaches demonstrated above may vary. A great deal depends on your own feelings of confidence in handling situations like this. Our view is that the first represents a good try at letting the client have her feelings acknowledged without being driven by the interviewer. However, you may wish to reflect on how the interview might be brought back to more legally relevant issues.

The second dialogue shows a sympathetic response, but it is one which tries to explain away the clients feelings by providing a possible rationale for the behaviour of her former colleagues. It appears to make matters worse. The solicitor then offers his client a cup of tea, enabling him to escape from the conversation for a while. His parting remark is a clear indication that he wants to get away from what he feels are the more irrelevant aspects of the interview.

The third dialogue is a clear indication that the interviewer doesn't regard what his client is saying about her feelings as important or relevant. This is assuming he has picked up the cues from what Mrs Tyler has said.

Active listening ensures that you tune into your client's concerns and anxieties as well as the information you need to give good advice. It enables your client to put things in their own words and enables you to help them identify their goals. Failure to listen actively makes it likely that your client will not be able to uncover all their concerns and consequently your advice may be flawed.

2.4.2.4 Using checklists

Some people suggest you prepare checklists to use during the interview. We urge you to use these with great care.

> Before the Interview with Mrs Tyler, you prepare a checklist of questions, knowing that she wants to tell you about an accident at work. Some of the questions you might have noted are:
>
> What's the problem?
> What happened?
> Where did the accident take place?
> When did it take place?
> Who is your employer?
> What is the injury?
> Had you received any training?
>
> These may all be legitimate questions to ask, but equally they may cause you problems during the interview. What problems could you envisage?

We can think of two problems with a checklist like this. First of all, you are assuming that Mrs Tyler is the victim of the accident at work and she therefore wants to bring an action in negligence against her employer. You are already thinking breach of duty and damage. You may well be correct, but you do not yet have enough information to be able to decide. It is possible, for example, that Mrs Tyler is an employer who would like to dismiss an employee who has caused an accident at work. Alternatively, she might be coming to see you because she thinks she is being victimised because of an accident she was involved in.

The second problem is that a list like this imposes an agenda on an interview which may prevent you from actively listening to your client and picking up cues from what she says and the way she says it. The danger is that you take a 'one question, one answer' approach and fail to follow up and funnel your questioning to probe the detail. For example, in an LPC assessment once, the interviewer was asked to handle the purchase of a small terraced house in a run-down part of town. Instead of listening carefully to the client, the interviewer, following his prepared checklist, asked: 'Are there any fishing rights attached to the property?'

Both of these problems may prevent you from helping your client to identify her concerns and objectives and reach appropriate decisions about how to proceed. Moreover, you are hardly likely to instil a sense of confidence in your abilities!

Nevertheless, inexperienced interviewers tell us they like to have a checklist with them to make sure they cover all the ground. Furthermore, they say it helps them keep control over the interaction. A skilled interviewer is able to listen carefully, think quickly and continuously modify their perception of the situation. This is a very sophisticated set of skills and many students lack confidence in demonstrating the level of skill needed to interview effectively. Consequently they feel that the more preparation they can do in the form of checklists, the more confident they are that they will have something to fall back on if they lose concentration or dry up.

We see interviewing principally as a listening, not a questioning activity. While we would not wish to completely discourage the use of checklists, we would suggest that rather than a list of questions, you frame them as a series of topics to be raised. In the past we have found it useful to write these topics not as a list, but in a circle. This means that you are not pre-empting or prioritising the order in which issues occur but you are making sure you cover all the issues you want to cover. It is also good practice to confirm with your client, as the interview proceeds, that the topics you have written down are in fact relevant. A topic list for Mrs Tyler's interview might look like this:

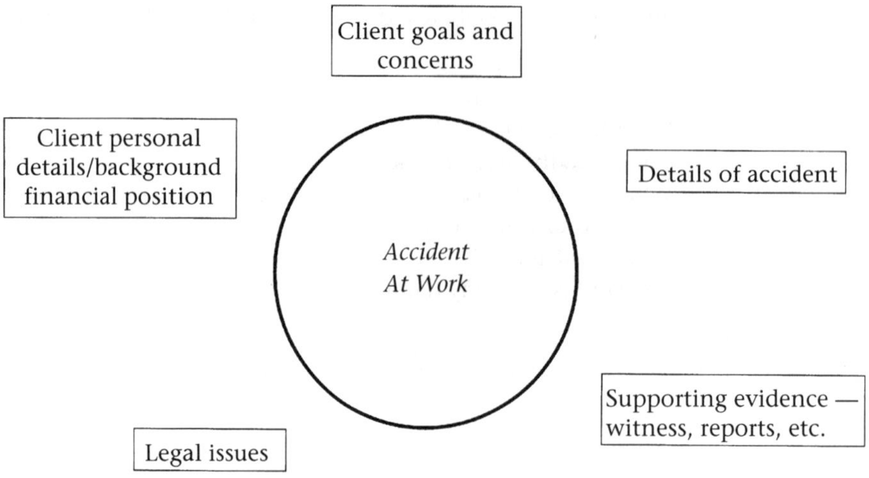

Not everyone is comfortable with this approach to the interview. However, we recommend that you try it out a few times. As you become more confident we think you will find it more useful. If you really don't like it, just bear in mind that any checklist you prepare must not have so much detail in it that during the interview you are over dependent on it and cease to listen and think!

To summarise the 'Acquire' stage of WASP, here are some Dos and Don'ts:

Dos and don'ts for acquiring information

Do	Don't
Use open questions and invitations to talk.	Be an interrogator.
Use funnelling techniques to get details.	Feel you have to break every silence.
Listen actively for facts, feelings and hidden agendas.	Assume you know what the problem is until you have all the information.
Summarise and reflect what you hear.	Ignore your client's feelings.
Acknowledge that your client may have strong emotions and be ready to discuss them.	Use detailed checklists.

We now go on to look at the part of the interview where you supply information and give advice. It is important to remember that you don't start to give advice until you have as full a picture as possible of your client's situation. This is where your skill and experience play their part. Clients may not be aware of what is legally relevant or they may be unwilling to give you certain information if, for example, it shows them to be dishonest or incompetent. It may well take more than one interview to get all the relevant information, but bear in mind that meetings with clients are costly and you begin to look less than competent if you have to keep arranging meetings with clients to ask questions you should have asked at the beginning.

2.4.3 Supply information and advise

> Note down all the types of information and advice you will give to your client in a first interview.

You may have thought of some of the following:

(a) whether the facts fit the law;
(b) the merits or otherwise of the case;
(c) what has to be proved;
(d) legal and non-legal options;
(e) the risks involved in proceeding;
(f) the costs;
(g) details of the services you provide;
(h) contact arrangements;
(i) what happens next and by when; and
(j) things the client has to do.

You may well have thought of other categories of information. However, when you are advising a client, the approach you take will not only depend on the information they give you, but it will also be influenced by your own perceptions of your role as a solicitor as well as by the expectations of your employer as to how clients are dealt with.

Various writers have identified a number of roles. Some seem to overlap. The point, however, is that you will probably develop a preferred role along with one or two other roles you can play when the occasion demands. Avrom Sherr (in *Client Care for Lawyers*, 2nd ed. (London: Sweet & Maxwell, 1999)) summarises the lawyer's roles as follows:

(a) The 'hired gun'—is highly adversarial and sees his task as getting the best possible deal for his client, regardless of the rights and wrongs of the dispute.
(b) The 'high priest'—hands down advice as if from on high. There is no sense of a collaboration between client and solicitor.
(c) The 'business person'—is concerned with handling cases efficiently and making a good living from his profession. This solicitor has little concern for the factual or emotional details of a case.
(d) The 'problem-solver'—sees himself as a professional whose job is to help the client achieve their objectives, whether legal or not. He will be interested in seeing what lies behind the facts of the case and in helping the 'whole' client.

> How do you think each of the sterotypes listed above would approach advising Mrs Tyler, given that you had obtained the following information from her?
>
> (a) Her employer has told her in writing that unless she is fit to resume work within two months, she will be retired on ground of ill health.
> (b) She is upset and angry about this.
> (c) She wants to go back to her job.
> (d) Her doctor thinks she may not be fit enough to go back to work in two months.
> (e) She is very upset that her relationship with her colleagues, especially Fran, seems to have broken down; she has talked at length about this and would like things to be as they were.
> (f) She has financial backing from her Trade Union.

The categories above are stereotypes, so we need to be careful in making predictions about how real people would behave in real situations. However, the lesson to be learned here is that your approach to giving advice is a negotiation between the culture of the firm which employs you, your own values about how you interact with your clients and your client's expectations about what solicitors do. It is not merely a technical question about getting the facts, applying the law to those facts and giving advice accordingly.

2.4.3.1 Giving legal advice at a first interview

In our experience novice interviewers tend to feel more anxious at this stage of the interview than at any other stage. There are a number of possible reasons for this. One is the perception that clients come to solicitors for expert legal advice and will question the interviewer's competence if they don't get it. One of the things they will certainly want to know is whether they have a strong case or not. Moreover, law students are not confident about the extent of their legal knowledge; they may not have 'covered' the legally relevant area, or if they have, may have forgotten all or most of the detail. It may even be the case, where facts and law are quite complex, that the interviewer is unable to immediately identify the areas of legal relevance. So what are they to tell the client, who is looking for answers?

As we said earlier, while the client is telling you their story you will be fitting the facts onto your cognitive legal frameworks—negligence, breach of contract, etc. Once you are satisfied that you have as full a version of the facts as possible, you should be able to give a preliminary opinion on the client's legal position. In many situations you won't be able to provide a real estimate of the merits or otherwise of the case until you have obtained further information (for example, witness statements, accident reports, medical reports) and researched the legally relevant areas in more detail. However, there are situations where you should be able to give more detailed advice. In circumstances where the client wants you to perform a specific task, such as deal with their house sale and purchase, or draft a will, you will be able to obtain sufficient information on which to base clear and detailed advice and to set out a plan of action.

EXERCISE 2.5

Based on the information Mrs Tyler has given you about her accident, what preliminary view have you formed as to her legal position?

(a) Work with a partner. One plays the part of Mrs Tyler, the other the interviewer. The interviewer advises Mrs Tyler on her legal position.

(b) Discuss what went well, what went less well and why.

(c) Swap roles and feed back to each other (and whole workshop).

You will probably have agreed that you need to give the client something meaningful so that they go away feeling confident about the situation and your handling of it. Consequently you will need to provide some information, even if it is only a preliminary evaluation of the merits of the case. In such a situation, you will also need to make clear that you cannot be more specific until you have investigated the facts further, e.g., seen the accident report and obtained an expert medical opinion. Also, you need to do further legal research before you can advise her with more certainty.

An issue that may have arisen during this exercise is how to communicate your legal advice in a form which the client will understand. Your academic law study will have prepared you to apply law to facts and estimate the strength of a fictitious client's legal position. However, your 'advice' will have been given to lawyers (your tutors or examiners) who share with you the language of law and legal reasoning. Unless you have participated in a clinical legal education programme or had some practical legal work experience you probably won't have advised a lay person who is not at all familiar with legal language.

EXERCISE 2.6

Look at the following scenarios.

(a) A client comes to you wanting a divorce and tells you that they want to marry the person they are living with as soon as possible. A friend has told them they have to wait for a 'decree nisi'. The client asks you what a decree nisi is.

(b) A client is very concerned that she is being threatened with a lawsuit for debt. It turns out that this debt was incurred by a former business partner who has now walked out on the business and has left the country. Your client no longer operates this business.
Explain to your client the implications of joint and several liability in this situation.

(c) Your client is defending a small claim brought by a supplier of computer hardware and software. This supplier had agreed to install a networked computer system in your client's offices, together with the most up to date and appropriate software. What was actually installed however, was an older, less powerful system with older software.
Moreover, the supplier never sent the user licences as promised. Your client has refused to pay the agreed price. Explain clearly to your client the implications of 'caveat emptor', 'satisfactory quality' and 'fitness for purpose' under the relevant legislation.

(d) You are called to the police station to advise a 12-year-old client. He is being interviewed about alleged repeated harassment of, and threatening behaviour towards, a particular group of families on a local housing estate. During the interview the interviewing officer says to your client, 'You're probably looking at an ASBO for this.'
'What's he on about?' Your client asks you.

(e) Your client is extremely distressed because she has been charged with intent to supply a class A drug. A friend had asked her to look after a small number of ecstasy tablets for a few days while his girlfriend's parents were staying at his flat. Driving home after taking possession of the drugs, your client was stopped by the police for speeding. They found the pills in a little plastic bag on the passenger seat of her car.
Your client says, 'I can understand I might get done for possession, but I wasn't going to sell them to anybody. I'm just keeping them for my friend. What's going on?'

Tasks:

1. Prepare your explanations to the clients.
2. Work with a partner. One plays the part of the client, the other the part of the lawyer. The lawyer gives their prepared answer to the client.
3. Swap roles and repeat the exercise.
4. What have you learned from this exercise about explaining legal technicalities to lay clients?

There are two rules to follow when explaining legal issues to a client:

(a) First of all, you need to explain legal matters in ordinary language. This is an important skill which you need to develop. It is of absolutely no use to a client receiving an explanation which they don't understand and which consequently explains nothing. Be careful not to overload your client with too much detail too quickly, particularly if it is not immediately relevant to their situation.

(b) Secondly, you must then relate your explanation to the client's own situation in such a way as to make it clear how the law sees their problem.

For example, in (a) above you would need to point out first of all that the client's friend had got it wrong! Following our rules, here is one possible explanation: when you get a divorce, there are two court orders issued which are called 'decrees'. The first one is a provisional order—the 'decree nisi'. When this order is made, you are not yet divorced, but you have a period of at least six weeks to make any final arrangements with your husband/wife about children and property. When all these arrangements have been made the court will make a final order—the 'decree absolute', which means you are officially divorced. In this client's case, they will not be free to remarry until the decree absolute is granted.

2.4.3.2 Helping the client decide what to do

In any situation there is always more than one course of action available to the client. It is your role to help the client decide which course of action is most likely to help them achieve their objectives. This means that you have to be clear about what courses of action are available from the information the client has given you. You need to set these out clearly and comprehensibly and spell out their consequences. Any course of action will have costs, benefits and risks attached to it. It is your job to ensure that the client fully understands what these are. Some of the options that you have set out may be contingent on getting further information and doing legal research. Once again, you must make this absolutely clear to the client. It is vitally important that you do not mislead clients into expecting outcomes which further investigation of facts and law will make impossible to achieve. Finally, it is the client's decision as to which course of action is followed. The client bears the cost and the risk and only they can decide the level of cost and risk they are able to tolerate.

2.4.3.3 Supplying other important information

At some stage during the interview you must give the client information about what it will cost them to instruct your firm and what services the firm provides. Not only are you expected to do this by the Law Society, but such information plays a material part in the client's decision whether or not to proceed, and if so, which course of action they would prefer to pursue. Views differ as to when clients should be given information about costs. One view is that general information should be given at the beginning of the interview. We believe there is room for flexibility. If a client is to be billed for the interview, they need to know what the charge will be right at the outset, so they can decide whether to continue or not. Ideally, this information should be given to clients when the interview is first arranged. As for the standard charges of your firm, we consider it good practice that they should be written down and supplied to clients as soon as the client decides to instruct you. However, clients should also be told about likely costs during the course of the interview, so that they can ask questions and clarify any misunderstandings.

It is just as important to tell the client what standards of client care they can expect from your firm. Therefore, you need to give them information on such things as how they can contact you, how regularly they will be updated on the progress of their case, how quickly you will deal with any queries or other communications from them. This information is important because it keeps the client informed and involved.

Dos and don'ts for supplying information and advising

Do	Don't
Set out all the options available to the client, once you have enough information.	Assume you have all the information you need from a first meeting.
Give clear information about costs.	Make the client's decisions for them.
Give clear information about standards of client service.	
Develop the skill of explaining legal issues in plain English for lay clients.	
Make sure that the legal explanations you give fit the client's circumstances.	

2.4.4 Part

The fourth part of the WASP process is 'Part'. It is important that at the end of an interview, your client has provided all relevant information and has been advised of their legal and non-legal options, the costs of proceeding, the risks they run by proceeding (or not proceeding) and what level of service they can expect from you. The last part of the interview process is to summarise and confirm all this.

> What needs to be done during the final stages of an interview?
> List what you think are the main things to deal with.

We think that you should finish an interview by:

(a) confirming that the client wishes to instruct you and understands what course of action is to be undertaken;

(b) confirming that the client is aware of the costs and risks involved;

(c) confirming that the client knows how to contact you or an appropriate colleague;

(d) confirming that the client is aware of any action they have to take and by when it needs to be taken;

(e) explaining that you will write to the client summarising these matters within a certain number of days; and

(f) checking that the client has no further matters or queries to raise with you.

Your client should leave with a clear understanding of what they have committed themselves to and a grasp of the legal issues which relate to their circumstances. They should feel satisfied that you will be helping them to realise their objectives and confident in your professional competence.

2.4.5 When to be flexible

The WASP approach that we have outlined above is designed to make sure that the important aspects of the meeting are dealt with: client care, getting relevant information and making sure that your client articulates their concerns. If your interview structure is too rigid, you may fail to do one or more of these things. So it is important that you know when to be flexible. Although a general principle of interviewing is 'do not start to advise until you have enough information', there may be occasions where you need to break this rule. For example, a client may come to you convinced that they are in serious trouble. It quickly becomes plain to you that they are overstating the problem. In such a case it would be a good idea to tell them this as soon as possible, to put their mind at rest.

Similarly, you will need to be flexible where you need to make an urgent bail application in court. In such a situation you have very little time to get the details you need from the client to make the application properly. In this case, your client will be better served by your getting straight to the point than by following the usual interview structure rigidly.

2.4.6 How good is your notetaking?

To be an effective interviewer you need to pay attention to your notetaking skills. Nothing is more off-putting to a client than the sudden scribbling down of something they have just said. On the other hand, you need to be able to recall accurately the details of the interview.

We suggest you develop the following techniques:

(a) Explain to your client at the beginning of the interview that you need to take notes in order to recall the details of the case for further action.

(b) Take a 'listen first' approach; this means that you begin by actively listening to what your client has to tell you. When you have checked your understanding of what they are telling you at each stage of the interview, then you can note it down.

(c) At the end of the interview, use your notes to summarise the information; this will enable you to check that your notes are an accurate representation of what took place.

(d) As soon as possible after the interview, write up your notes clearly and accurately. It is important to do this because notes taken at the time may serve your short-term memory well, but may cause you problems if you rely on them days or even weeks later.

(e) Keeping accurate notes is a very important part of your work. You will rely on them to supply information on which further action may be based. If they are not accurate, you may make mistakes or irritate your clients by having to go back to them to ask again for information they have already given you.

2.5 Interviewing and advising: an exercise

2.5.1 Planning an interview

This is an opportunity for you to plan an initial interview with a professional client.

Your client's name is Joanna Lee. During a brief telephone call she gave you the following information.

She is currently a self-employed computer systems analyst, earning on average between £40 and £50,000 a year. She has been headhunted for and offered a senior management job in a Bristol-based company, Avon Systems International, at a salary of £58,000 plus benefits. She is presently deciding whether to take the job. The details of the post are provided in the draft employment contract which she has been sent to sign.

Ms Lee is happy with most of the package on offer but there are some issues raised by the contract which she is concerned about. Her main concern is the breadth of clauses 11.4 to 11.7. She is worried that these seem wide enough to prevent her leaving the company and working in the field for at least 12 months. She wants to know if the company is within its rights to demand this.

Her other main concern with the contract is clause 11.2. She has already done a lot of freelance software development work and holds copyright in the applications she has designed. The company want her to bring this software with her into the company and she wants to know whether copyright will transfer to the company under clause 11.2. At the job interview, she was told that in return for ceding copyright to the company, she would be given a share option in the company worth £20,000.

She is also being headhunted by another company, based in Manchester. Although the Manchester firm is in the same business as Avon, they have a much smaller market share, and are keen to employ Ms Lee, as her expertise would be a good fit with their medium-term strategy. They may be prepared to top Avon's financial package to get her, but there is an element of uncertainty about their long-term future. They have not yet interviewed her.

Ms Lee has sent you a copy of the draft contract for you to familiarise yourself with the relevant clauses. These are:

11.2 The copyright in all articles, designs, drawings, programmes, calculations, specifications, photographs and similar documents and written material produced by you in the course of your duties with the Company shall belong to the Company and such matters and copies thereof in your possession shall be returned to the Company on termination of your employment.

11.4 You shall not during the period of 12 months after the Termination date, directly or indirectly, either on your own account or otherwise, canvas or solicit business from any Customer Connection with whom you have had any material dealings in the Contact Period in the course of the employment.

11.5 You shall not during the period of 12 months after the Termination Date either on your own account or otherwise, do business with any Customer Connection with whom you have had material dealings in the Contact Period in the course of the employment.

11.6 You shall not during the period of 12 months after the Termination Date, directly or indirectly, induce or seek to induce any Skilled Employee, with whom you have had material dealings in the course of your duties hereunder in the Contact Period, to leave the Company's Employment whether or not this would be a breach of contract on the part of such employee or offer employment or an engagement to any such employee.

11.7 You shall not during the period of 12 months after the Termination Date, carry on or be interested in Competitive Business in competition with the Company for Access Accounting Systems whether as principal, agent, owner, director, employee or otherwise.

11.8 Each of the restrictions contained in this section 11 are considered reasonable by the Contracting parties being no greater than is required for the protection of the goodwill of the business of the Company and are intended to be separate and severable.

Ms Lee has made an appointment to see you the day after tomorrow. Using the WASP structure, prepare your interview plan:

(a) What are your objectives?

(b) What do you see as the important information so far?

(c) How will you verify this? and

(d) What areas of law do you need to research as far as you can tell at the moment?

When you have carried out these tasks, compare your answers with those of a colleague. What are the similarities? What are the differences?

2.6 Learning outcomes

In this chapter we looked at the basic skills and processes involved in interviewing and advising clients. You should now be able to:

- state the purpose of the client interview;
- use all information-seeking techniques confidently;
- listen actively and respond with appropriate follow-up to what you have heard;
- summarise information to check understanding;
- structure an interview so that necessary information is acquired and given;
- give appropriate advice to the client;
- state the processes whereby rapport with clients can be developed.

2.6.1 Law Society written standards

Throughout your career as a solicitor you will be constantly reviewing and improving your skills. The Law Society provides standards of competence which we have reproduced below. These standards will be used to assess your skills during your training. As your career progresses you should return to the standards and check your own skills against them.

Students should be able to identify the client's goals, gather information to identify means of realising those goals and assist the client to reach decisions as to the appropriate means of implementing those goals.

Students should be able to:

(1) prepare for the interview;
(2) deal with appropriate client care and professional conduct issues during and after the interview;
(3) allow the client to explain his or her concerns;
(4) identify the client's goals;
(5) help the client reach priorities among those goals;
(6) elicit relevant information and distinguish between relevant and irrelevant information;
(7) use appropriate listening and questioning techniques;
(8) determine what further information is required;
(9) identify possible courses of action and the legal and non-legal consequences of selecting a particular course of action;
(10) assist the client to make a decision regarding the best course of action, including the costs, benefits and risks of that course of action;
(11) agree the action to be taken by both parties subsequent to the interview and an appropriate time-frame for such action;
(12) accurately record the interview, confirm instructions and confirm the action that needs to be undertaken;
(13) establish a professional relationship with the client and deal with any client care or professional conduct issues that may arise when advising the client.

Students should be able to demonstrate an understanding of the principles and criteria of good interviewing and should be able to:

(1) conduct an interview in which those criteria are met;

(2) modify their own interviewing errors to meet these criteria;

(3) transfer skills learned in one context to other contexts;

(4) develop techniques for appraising and developing their own interviewing style.

Students should be aware of the need to deal with a range of personnel including other lawyers, professionals, witnesses and parties. Students should be aware of the need to take instructions from their principal and other supervising solicitors.

2.7 Self-test questions

(1) Describe five techniques for encouraging your client to talk.

(2) What is active listening?

(3) Why is non-verbal communication important during an interview?

(4) What does the client need to know at the conclusion of the interview?

2.8 FURTHER READING

Bolton, R., *People Skills* (Brookvale, New South Wales: Simon & Schuster, 1987).

Maughan, C., & Webb, J., *Lawyering Skills and the Legal Process*, 2nd ed. (Cambridge: Cambridge University Press, 2004), Chapters 3 & 5.

Sherr, A., *Client Care for Lawyers*, 2nd ed. (London: Sweet & Maxwell, 1999).

3 Legal writing

3.1 Learning outcomes

This chapter deals with the skills needed for effective legal writing. In this chapter we focus particularly on:

- the 'Plan, Write, Revise' approach to effective writing;
- strategies for developing your ability to write plain English;
- how to vary your language and style to suit the needs of your reader;
- being self-critical and continuing to develop your writing skills.

3.2 Introduction

You have all been writing for 15 years or more, and you have all been highly educated in a system in which writing played a crucial part. Why then, do you need to work through this chapter?

We think there are several possible reasons:

(a) You will probably have done most of your writing in school, college and/or university, and while that seems like quite a lot of writing, it is different from the kind of writing you will need to be a competent legal practitioner.

(b) Have you really had many years' writing experience, or have you had a year's experience repeated many times? In other words, have you learned from your experience, or have you just reinforced bad habits?

(c) You may not have a clear idea about what makes effective writing. You need this in order to be able to vary how you write, depending on what you write and who you are writing to.

(d) Your recent experience of writing will have been as a law student. You will have written essays and answers to problems for other lawyers to read and evaluate. This means that you will have got used to the discourse of law (the language and method of reasoning used by lawyers) and will probably use it automatically in your legal writing. As a practising solicitor, however, much of your writing will be aimed at lay people.

The purpose of this chapter is to enable you to become a competent writer in professional practice.

3.3 Why write?

Why did human beings invent writing? Consider the following documents:

(a) a will;

(b) a TV licence;

(c) a letter to a client on holiday abroad, informing her of the date of her court appearance;

(d) a degree certificate;

(e) Cheshire & Fifoot's *Law of Contract*;

(f) an inventory;

(g) a diary entry;

(h) an arrest warrant.

> Why do you think we need to have these documents in writing? Try to think of at least two purposes of writing and note them down.

You may have thought of others, but here are the reasons we came up with:

(a) Writing enables you to communicate accurately through time; what you write can be stored and read later, either by you or others.

(b) Writing enables you to communicate at a distance.

(c) Documents can be published, and so read by large numbers of people.

(d) A written document may authenticate certain actions by being signed, witnessed, stamped, or marked in some other way, e.g., a will, certificate or licence.

Note that a will fulfils at least three of these purposes; it is communicated over time in that it is written by someone who will never see it enacted. It is authenticated by being signed and witnessed, and is stored in a public records office for anyone to see.

These are the reasons why people write, but the purpose of writing these documents will only be fulfilled if their meaning is clear to those who read them. The information must be accurate and clear; writers may not get a second chance to communicate their meanings in a way that they want them to be communicated.

Lawyers need to communicate through time, over distance and to create permanent records. But the fact that their education and training immerses them in 'legal language' often leads them to forget that many of the people they write to find such language very difficult to understand. The effect of this has been to distance the law from the public and lawyers from their clients. Moreover, this distance between lawyer and client can lead to poor relationships, resentment and a feeling of powerlessness on the client's part, whilst lawyers may feel frustrated that they are unable to get their message over.

3.4 Know your reader

One of the main aims of this chapter is to make you aware that different readers require different approaches from writers.

> Imagine you need to convey the same information about an industrial injury to three separate people:
>
> (a) your client, the claimant;
>
> (b) the defendant's solicitor;
>
> (c) the client's doctor.
>
> Consider how you would communicate this information in the three different cases.

Each of these recipients is an individual with whom it is important to establish and maintain a fruitful relationship. You will therefore need to vary your language and style not only to suit the status of the recipient, but to show your awareness of their individual requirements. The way you write will be determined by a number of factors:

(a) how well you know the person. This will determine how formal you need to be: how do you address them and what kind of language do you use?

(b) how easy they find it to grasp the issues involved. How much explanation will you have to give if the information is complex, detailed or technical?

(c) their likely attitude towards the message. Is the news good or bad? What is the likely response from your reader?

(d) their attitude towards you, the writer. Do you have an ongoing relationship? Is your reader a fellow-professional? Is the reader on your side or the other side?

(e) their reading ability;

(f) their understanding of English, if English is not their mother tongue;

(g) what the outcome is likely to be for them.

EXERCISE 3.1

(a) To practise varying your style, try writing the following for a non-lawyer:

(i) Your sister phones to say she has been arrested for stealing two bottles of vodka from a supermarket. You start to explain the law on theft to her over the phone, but get cut off. Fax her an explanation.

(ii) A friend of your father runs a business developing computer software. He has reason to believe that a rival organisation intends to sell his applications and wants to know what emergency measures could be taken to prevent this. He wants you to e-mail him your reply. You want to suggest the use of a search order. How would you explain your recommendation to him?

(iii) A friend e-mails you to say he has had his wrist broken in a rugby match. He wants to know if it is worth his while to claim compensation. E-mail him to explain the volenti non fit iniuria principle.

(b) Work with a partner. Compare your written versions.

(c) Discuss the methods and techniques you used to explain the legal terms.

(d) Discuss and note down what you have learned from the exercise.

(e) If you were asked to repeat this exercise, would you make any changes to your approach? If so, what changes would you make?

It will have become obvious from this exercise that in order to get your meaning across effectively you need to be clear about the meaning yourself. Because we are familiar with and constantly use technical language and jargon we no longer need to think about its meaning. It then becomes difficult to express the meaning in everyday language.

The ability to write effectively begins with an appreciation of the needs of your reader. Remember, if what you read is confused or difficult to understand, you have wasted your time writing it. This makes you look unprofessional and is a poor recommendation of your firm. Effective writing is not about pleasing you, it is about pleasing your reader.

3.5 Strategies for effective legal writing

3.5.1 Plan, write, revise

There is no one best way of tackling a piece of writing. People use a variety of approaches. Sometimes the same individual uses different ways of dealing with different types of writing. You need to experiment with the way you find best for you. And remember, if your way works, you don't need to feel guilty that you are not following some pre-determined recipe recommended by the experts.

We are going to recommend the approach of *Plan, Write, Revise*. There are several reasons for this:

(a) Time is money for a practising lawyer. You cannot afford to spend time on repeated rewriting and revising material which has been poorly planned in the first place.

(b) Plan, Write, Revise is a logical approach that fits well with the problem-solving nature of legal work.

(c) It is a straightforward strategy that is easily grasped by those of you who may not have had a very methodical approach to writing in the past.

(d) As you become a more experienced lawyer, you will become increasingly familiar with the matters you need to write about. You will also build up relationships with clients and other professionals. You will not therefore have to expend the same amount of effort over every document that you write as you do now. There is a danger that you may become too complacent about your writing ability. If you have internalised the Plan, Write, Revise technique, good written communication habits will have become automatic.

3.5.2 Plain English

Martin Cutts describes plain English as:

> The writing and setting out of essential information in a way that gives a co-operative, motivated person a good chance of understanding the document at first reading, and in the same sense that the writer meant it to be understood.
>
> Cutts, M., *Plain English Guide* (Oxford: Oxford University Press, 1999).

Campaigns to promote 'plain' speaking and writing developed in Britain and the USA in the late 1970s as a method of attacking the over-complicated language used by government, local authorities and businesses to communicate with the public. Their aim is to replace this 'gobbledegook' and 'babblegab' with clearer forms of expression. Well-known examples include 'collateral damage' (civilians killed in a war) and 'negative patient care outcome' (death).

However, as Cutts points out, support for plain English speaking and writing is not new. For example, in 1550 Edward VI remarked: 'I wish that the superfluous and tedious statutes were made more plain and short, to the intent that men might better understand them.'

Opposition to plain English documents has come largely from the legal profession, which fears the risk of ambiguity and increased litigation they foresee could result from movement away from the tried and tested legal formulations. Nevertheless, lawyers now accept that simplifying grammatical structure, adding punctuation and eliminating archaic and Latin expressions will save them time and money and simplify their job because their documents become more intelligible.

> Can you think of any areas of legal practice where traditional terminology is being discouraged, or even eliminated?

Changes to the Civil Procedure Rules in 1998 were intended to shift attitudes towards openness and fairness and away from adversarialism, where each party attempts to score points over the opponent. A move towards plain English is seen as part of this shift.

For example, under the new rules 'plaintiffs' became 'claimants'; 'pleadings' became 'statements of case'; Mareva injunctions became 'freezing orders'; and Anton Piller orders became 'search orders'. This was certainly a start, but it is debatable whether some of the CPR terminology is sufficiently clear to avoid confusion.

You may have identified other examples. One is the requirement under the Unfair Terms in Consumer Contracts Regulations 1999 that a seller or supplier should ensure that any written term of a contract is expressed in plain, intelligible language (see Reg. 7(1)).

3.5.3 Planning: know what you mean to say

To ensure that your message is clear and precise, you must know exactly what you mean to say and how you mean to say it to this particular recipient.

Begin by clarifying the purpose of your communication. Are you merely giving information, or are you attempting to persuade someone to take a particular course of action, or are you responding to a complaint? Decide what you want to happen as a result of what you write.

Think carefully about the content. Do you have all the information you need? Is it accurate? Remember, others will rely on what you say.

Decide the most logical and appropriate structure for your content. You will normally be doing one or more of the following in your writing to clients:

(a) answering their enquiries and questions;

(b) giving detailed advice and information;

(c) setting out options; or

(d) giving advice on appropriate courses of action.

The main thing to remember is that your client is not your law teacher! They are not interested in the process by which you reach your opinions about their case. They are interested in your advice being consistent and accurate, and in its being given to them clearly and unambiguously.

The general principle is to get straight to the point. Your client doesn't want to wade through paragraphs of legal reasoning before finding out what they really want to know (if they can find it at all). So the rule is: give the answer first, then say why it is the answer. This is the opposite to what you have been trained to do in law school, of course, where your tutors are much more interested in seeing how you reason to a conclusion than in just seeing what that conclusion is.

Of course, there are times when there will not be a clear-cut answer. In such a case the principle is similar: set out each option in turn followed by its attendant risks and costs.

Remember that each paragraph should deal with one main topic. The most common structure for a paragraph is to begin with a word or phrase linking it to the previous one and then to go on to introduce the new topic. The rest of the paragraph explains, illustrates, modifies or otherwise develops this topic. We deal with paragraphing in detail at **3.5.4.7**.

Moreover, you should vary the length of your paragraphs to add variety and interest, just as you would vary the sentence length.

Consider how your writing will look on the page. There is nothing more daunting for the reader than closely spaced, undivided blocks of print. If you are providing a lot of information, use headings, sub-headings and a numbering system to break up the information on the page and to indicate the relationship between various pieces of information.

To summarise, planning involves:

(a) knowing your reader;

(b) knowing your purpose;

(c) knowing your content;

(d) knowing your structure; and

(e) deciding your layout.

3.5.4 Writing: say what you mean to say

No doubt you are aware that there are differences between the spoken and written language. Make a note at this point of what you think those main differences are.

You may have noted that the spoken language has a number of features that are absent from the written. If you watch people talking you will see that not only do they use the vocabulary and grammar of the language to communicate their meaning, they also convey meaning by using body language (gesture, posture, facial expression). Body language tells each of the participants the state of mind and level of understanding and interest of the others. Moreover, you will hear the speakers emphasise certain points in what they say by the tone of their voice. Speakers and listeners utter 'wordless' sounds from time to time, for example, 'ugh', 'um', 'phew', which convey disgust, agreement, sympathy, the need for time to think, and so on. It is this combination of body language, intonation and wordless sounds that makes the spoken language so much easier to understand than the written.

Furthermore, a conversation allows the participants to respond immediately to what is said, and it allows for modification of what is said in the light of that response. With writing there is no immediate response, so you need to get it right first time. It is therefore hardly surprising that competent writing is thought to be such a difficult skill to master.

Another difference you may have noted is that in writing we are tempted to use words and phrases which we would use less frequently in speech: 'commence' for 'begin', for example, or 'in the event of' for 'if', 'for the duration of' for 'during'. We think that people use these forms because they believe they add a veneer of sophistication and importance to their writing. Unfortunately, adherence to this view is probably one of the main sources of much of today's obscure or unintelligible 'officialese'! Here are some examples:

> Prior to the collection of your vehicle please ensure that you pay for your parking at the automatic machines located at the administration building which is situated at the exit.
>
> I have given implicit instructions to my staff to keep noise to an absolute minimum due to the close proximity of residential properties.
>
> We have been requested by your pension company to acquire confirmation in writing from yourself that all the necessary requirements have been complied with.

Since clarity is your main aim, try to avoid the appearance of pomposity and unnatural, unnecessary formality which infrequent words and phrases give to writing. While the addition of the occasional infrequent word may do nobody any harm, the gradual piling up of such words and phrases in a piece of writing will make it turgid, dull and hard work for the reader. For the same reasons you should avoid long-windedness, unnecessary jargon and overlong sentences.

Criticism of lawyers' communication skills is not new. In 1702 Daniel Defoe wrote that the English of the clergy, physicians, academics and lawyers 'has been far from Polite, full of Stiffness and Affectation, hard Words, and long unusual coupling of Syllables and Sentences' (*Essays Upon Several Subjects: or, Effectual Ways for Advancing the Interest of the Nation*). More recently in 1963, David Mellinkoff described four linguistic mannerisms characteristic of legal writing: wordiness, pomposity, dullness and lack of clarity. (*The Language of Law*, 1963.)

As we noted earlier, competent written communication involves recognising and anticipating the needs of your reader. This involves more than explaining complex legal terminology to the lay person, which you attempted to do at **3.4**. The degree of 'stiffness' you select for your communication may determine whether your relationship prospers or not. You need to select language forms (grammar and vocabulary) which not only make the meaning clear as rapidly and straightforwardly as possible but which do so with an appropriate degree of formality.

We suggest you adopt the Three Cs as your starting point. Be:

(a) clear;

(b) concise;

(c) correct.

This means:

(a) Use frequently used words and phrases instead of infrequently used ones (see **3.5.4** above).

(b) Avoid clumsy and inelegant words and phrases.

(c) Use verbs in preference to nouns.

(d) Use active verbs instead of passive.

(e) Omit redundant words and phrases.

(f) Use specific words in preference to vague words and phrases.

(g) Avoid using jargon and technical terms unless there is no alternative and the reader will understand them.

(h) Avoid long and complex sentences.

3.5.4.1 Inelegant writing

Some formerly infrequent usages have now become commonly used in both speech and writing. Writers pad out their communications with additional words to avoid using a single word. The following are common examples:

- at a later date (later);
- until such time as (until);
- in the event that (if);
- prior to (before);
- subsequent to (after);
- on a regular basis (regularly);
- as a consequence of (because);
- due to the fact that (as, because);
- in the vicinity of (near);
- for the purpose of (to).

In letter-writing, this padding device is often used to link paragraphs, e.g., 'So far as this matter is concerned, we will deal with it as a matter of urgency'. We would prefer something like: 'We will deal with this matter urgently'.

We prefer it because it is short, clear and makes an economical link with the previous idea by using the phrase 'this matter'.

Clarify your thoughts when planning your communication so as to link your paragraphs concisely and avoid these clumsy devices. Using concise language will bring elegance to your writing.

3.5.4.2 Prefer verbs to nouns

A common feature of officialese and 'legalese' is the use of nouns instead of verbs when expressing actions. Again, we think writers do this to lend a false sense of sophistication and importance to their work. Generally, though, you will find that what you write comes over with much more simplicity and power if you use verbs. Furthermore, verbs bring vitality to your writing. Verbs convey actions, events, people doing things: living, breathing, eating, etc. Remember that law is part of the real world; claimants complain, judges decide, defendants pay. We hope you will agree that this sounds more real than: make a complaint, make a decision, or make payment. As Richard Wydick says (*Plain English for Lawyers* (Carolina Academic Press, 1985)), base verbs are 'simple creatures who cannot tolerate adornment. If you try to dress them up, you squash their life and motion'.

A base verb that has been turned into a noun is called a 'nominalisation'. Here are some more examples:

- file a complaint;
- draw conclusions;
- make a statement;
- make an objection;
- make provision for;
- take into consideration;
- make an amendment.

Additional words can, in Wydick's words, 'swarm like gnats' around nominalisations. For example, they can be expanded with adjectives: 'make major detailed provision for'; 'draw logical, irrevocable conclusions from'...

They can sound especially cumbersome, verbose and lifeless when expanded with the preposition 'of': 'the date of the commencement of the contract'; 'The administrative procedure will commence on the date of submission of the application'.

These expanded noun phrases are common in newspaper headlines to give readers as much information as possible in as few words as possible: 'Dead woman car boot sale puzzle'.

To add vigour and interest to your writing, stick to verb forms where you can.

3.5.4.3 Active v passive

Use of the passive is one of the most common reasons for excessive formality in writing, because it concentrates on what happened, treating the people involved as less important. Here is an example:

> When you arrive at the court you *will be met* by my clerk. You will be taken to court 2 where the case *will be heard* by Judge Jeffries.

This may be acceptable if you have a very formal relationship with your client, but the effect is to de-personalise the communication. It would be more direct and friendly (and probably more reassuring) to write:

> My clerk, Brian, *will meet* you when you arrive and *take you* to court 2. Judge Jeffries *will be hearing* the case.

In many modern word processors the grammar and style checker will highlight passive phrases like the ones above, and suggest that you change them to active. In fact, when we were writing this section, the style checker kept insisting that we should consider altering the passive to active sentences!

The passive is fine where there is no known subject, or where the subject of an active sentence would be very vague and meaningless.

> Which of these do you prefer and why?
>
> *Passive*
> John Smith was run over on 16 October. The driver of the vehicle has not been found.
>
> or
>
> *Active*
> Someone ran John Smith over on 16 October. No-one has found the driver of the vehicle.

In most English sentences the important information tends to come before the less important information. The important information in this communication is: 'John Smith was run over. We don't know who did it.'

The passive sentences put this information near the beginning, whereas in the active sentences the message is diluted because we have to invent vague subjects like 'someone' and 'no-one'.

In 'officialese', however, the subject is often omitted for less satisfactory reasons. It gives the impression that a sequence of actions takes place without any human intervention causing them, or seeing them to a conclusion. No wonder clients and other members of the public who are on the receiving end of this kind of writing see it as cold, remote and inhuman.

> Look at the two extracts below. The first is from the UK Passport Application form, the second from an information leaflet which accompanies it. We have italicised the passives, but you might also note other barriers to clear communication. What impact do they have on the reader, and how might they be redrafted to have a friendlier and more positive impact?
>
> Information *contained* in this form and on the passport record to which this application form relates *may be passed* to other government organisations and law enforcement agencies for the purpose of checking your application and in the subsequent use of any passport *issued* as a result of this application.
>
> Please note that if during the validity of a passport on which a child is *included* a new passport is *required*, for reasons of loss or amendment, a separate passport for the child *would need to be issued*.

The use of the passive distances the writer from the reader and makes it more difficult to understand on a first reading. It also makes the text seem very impersonal and unfriendly. In the first example, the last phrase 'and in the subsequent use of any passport issued as a result of this application' is frankly unintelligible because the sentence structure doesn't make clear what *and* links with in the earlier part of the sentence.

In the second example, the meaning is very simple: 'If your current passport has a child on it and needs renewing, then we must issue a separate passport for the child.' So why did the writer feel it necessary to use such complex grammar?

3.5.4.4 Redundancy

Redundancy is an important part of the language. It conveys the same message in more than one way so as to ensure understanding. For example, when you ask a question, the fact that it is a question may be conveyed:

(a) by the use of a question word;

(b) by a change in word order; and

(c) by intonation, or in writing by punctuation:
What time shall we meet tonight?

The redundancy that is built into language in this way is unavoidable and useful. However, the kind of redundancy that is not acceptable is the insertion of unnecessary words and phrases. These have the effect of padding out your writing, reducing its clarity and conciseness and diluting the force of the message. Here are some examples:

- *forward* planning;
- *close* proximity;
- *duly* incorporated;
- I enclose *herewith*;
- *tried* and tested.

3.5.4.5 Use specific words and phrases

It is unhelpful to clients to give them information which is vague or unclear.

> What is vague or unhelpful about the following?
>
> (a) You could be awarded substantial damages.
>
> (b) Getting the information we need should not take too long.
>
> (c) I will contact you in the near future.

In the first sentence, the client doesn't know what 'substantial' means. They are probably unaware of the level of compensation awarded by courts in different types of case and the word tells them very little.

In the second and third sentences, 'take too long' and 'in the near future' are not specific enough to be of any help to the client. You might as well not bother to write these sentences at all. If you put yourself in the client's shoes, you will realise that you need to give as much specific detail as you can. Where you can't be specific, for instance about timescales and damages awards, you should give an estimated range based on your knowledge and experience.

These points may seem obvious but are frequently forgotten by inexperienced students and trainees in their unwillingness to commit themselves to something more definite.

3.5.4.6 Jargon and technical terms

All occupations and professions have words and phrases which the members of those groups understand. As we noted earlier, outsiders often have difficulty with this and it can act as a barrier to good communication. Lawyers have often been accused of mystifying the public with their insensitive use of jargon and technical legal terms. Consider this example from a letter to a client:

> So far as your house purchase is concerned I enclose herewith a copy of a Deed of Grant dated 1967 which conveys to the Water Company in fee simple a right of easement to enter in and upon the lands.

The solicitor will have lost the reader long before any explanation is given, if indeed any is forthcoming.

Here is a list of terms lawyers are in the habit of using which you can safely exclude from your vocabulary:

- aforementioned;
- aforesaid;
- hereby;
- heretofore;
- herewith;
- inter alia;
- forthwith;
- pursuant to.

Feel free to continue adding to this list throughout your legal career.

3.5.4.7 Paragraphing

This is the skill of constructing documents whose contents flow in a logical order. We reminded you in **3.5.3** that one paragraph should deal with one main topic, which is then developed in the rest of the paragraph. Each paragraph should be linked in some way to the previous one and the following one.

Long, unbroken blocks containing large amounts of information are a sign of poor writing. They put a severe strain on the reader's information processing abilities by making them struggle to extract the information and make connections between pieces of information.

Almost as bad are paragraphs formed of just one sentence, because the writing seems disjointed and simplistic. (Like this one?)

Each paragraph has, at or near the beginning, a topic sentence, which encapsulates the main point of the paragraph. If you have planned your content carefully, you should not find it hard to decide what goes into your topic sentence. What is much more difficult, however, is relating the other sentences in the paragraph to the topic sentence and to each other. If you analyse a paragraph in detail, you will find that sentences relate to each other at various levels. Here is an example. We have given each sentence a number, in brackets:

The trouble with—and glory of—language is that it won't (indeed can't) be bound by rules (1). They do exist but not like standard-gauge railways (2). Sometimes we may want to write ain't and not is not or even isn't, to make a point or produce an effect that lies beyond the 'regular' standard (3). There are also many areas where people take sides or can't see what the problem is—like avoiding generic 'he' or splitting infinitives (4). Nobody can lay down absolute laws about these things (5). Nobody. No government. No council. No teacher. No guru of grammar or style (6).

(Taken from T. McArthur: 'Language Used as a Loaded Gun', *The Guardian*, 20 April 1993.)

The tree diagram below illustrates the relationships between the sentences. All sentences ultimately relate back to the topic sentence (TS). Level 2 (L2) sentences are subordinate to TS, but are equal with each other. Level 3 (L3) sentences are subordinate to their L2 sentences.

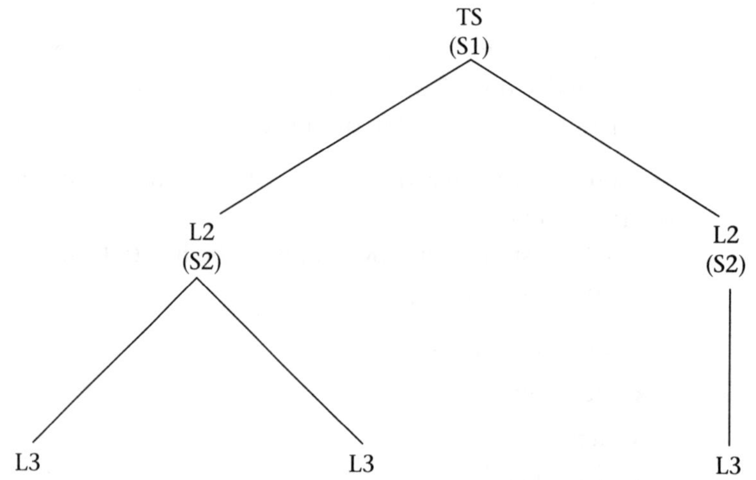

What the diagram means:

(a) *Sentence 1: TS*
This is the main idea of the paragraph. All other sentences develop it in some way. Stated simply:

'Language is not bound by rules.'

(b) *Sentence 2: L2*
This modifies TS:

'This is not to say there aren't any rules. There are, but they are not fixed or standard.'

(c) *Sentences 3 and 4: L3*
These illustrate the point made in L2 by showing that the rules are broken for a variety of reasons:

'For example, people use non-standard forms, or differ on what the rule should be, or do not recognise the problem.'

(d) *Sentence 5: L2*
This modifies TS.

'There is therefore no point in prescribing rules. (Why not? Because language won't be subject to them—TS.)'

(e) *The remaining 'sentences': L3*
These emphasise the point made in L2 above.

It is possible to analyse this paragraph differently, for example, by treating Sentence 4 as the topic sentence, or Sentence 5 as subordinate to Sentence 4. Your interpretation of the paragraph may be quite different from someone else's, or from the interpretation the writer intended.

So what have you learnt from this exercise? You should have discovered that paragraph analysis is a complicated business. It forces you to extract the writer's meaning by working out the development of his theme clearly and logically. This can make demands on the intellect, even where the writer writes clearly and coherently, as in the paragraph we have analysed. It is a much more difficult task where the writer's meaning is obscure.

When planning and composing, it is not only necessary to relate thoughts logically to each other within paragraphs. You must also link your paragraphs so as to give your document an overall coherent structure.

EXERCISE 3.2

The following text is taken from **4.8.4**, Your Draft in Their Hands. Read it carefully and answer the questions. If possible, work with a partner and compare answers.

Your client

As has been emphasised before, a solicitor produces documentation at the request and expense of the client. It is therefore sensible to send a draft of an agreement to the client for comment. It is not for the client to check that you have got the agreement right; it is expected that you will have the law correct. This is the opportunity for the client to give you the missing pieces of information that may not have surfaced at the interview, or for you to ask the client what is to occur in an eventuality to which you have not previously drawn the client's attention.

How this process occurs often depends on the client and the nature of the documentation. If the client is sophisticated, used to reading documents drawn by your firm and familiar with the type of transaction, it is more sensible to end the draft with a few pointed comments. On the other hand, if the document and/or the legal language is likely to be unclear, it may be better to discuss the document in a meeting with clients.

Of course, not every document needs to be checked with the client before submission to 'the other side'. It would be unnecessary, for example, for the client to see a standard conveyance. On the other hand, the client should always have an opportunity to see and comment on a draft will.

(a) Identify the topic sentence in each paragraph.

(b) Taking each paragraph in turn, identify the relationships of sentences to each other (TS, L2, etc). If it helps you, use a tree diagram.

(c) Taking each paragraph in turn, identify the words and phrases which:
 (i) link sentences within paragraphs, and
 (ii) link paragraphs.

 For example: 'and', 'or', 'but', 'however', 'moreover', 'therefore', 'on the other hand', 'for example', 'although'.

(d) Are there any sentences which are not linked by such words and phrases? If so, how are they linked?

As we noted earlier, paragraphs can be analysed for meaning with differing results. Here is our analysis. Yours may be slightly different.

Paragraph 1

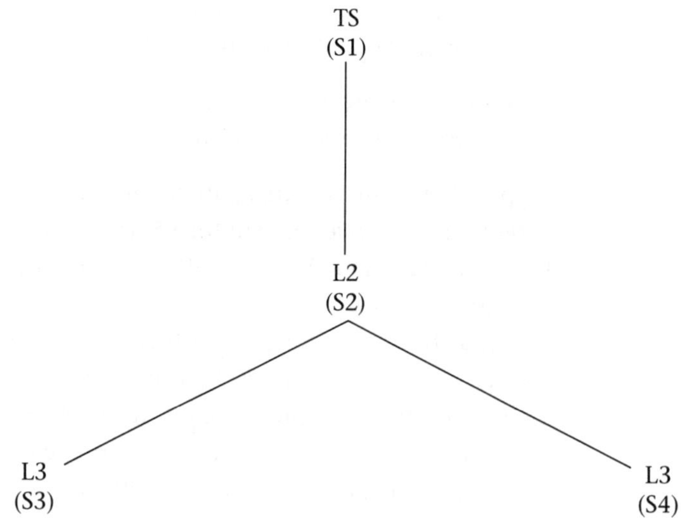

Paragraph 2

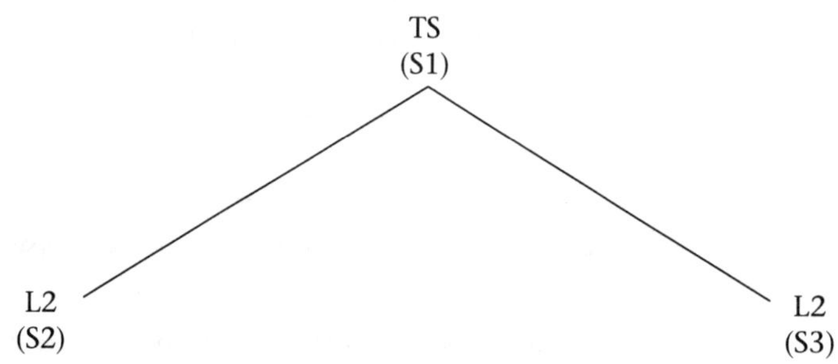

Paragraph 3

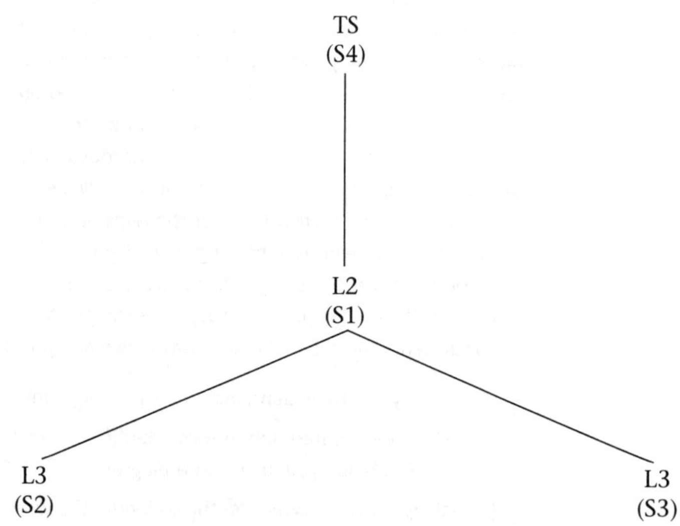

You have probably identified the following linking devices:

(a) *Paragraph 1*:

S1: 'As has been emphasised before, . . . Expenses of the client.'

This makes a connection with previous text, and so helps the document flow. The whole sentence is a link with what has gone before, and is the TS.

S2: 'therefore'—develops the topic in the TS.

S3: no explicit link word or phrase. The link is implicit, made by a repetition of the word 'client' in the same grammatical form as it is in the previous sentence (i.e., in S2 'for the client to comment on', and in S3 'not for the client to check'.)

S4: 'This... Or'. This sentence has two parts, linked by 'or'. 'This' refers back to S2—the sending of the draft.

(b) *Paragraph 2*:

S1: 'This (process)' links paragraph 2 to 1. This is the TS.

S2: no explicit link word or phrase. Link is repetition of 'the client' from S1.

S3: 'on the other hand' provides a contrasting example to S2 and refers back to S1.

(c) *Paragraph 3*:

S1: 'Of course', links paragraph 3 to 2 but is *not* the TS.

S2: 'For example', links with S1.

S3: 'On the other hand', also links with S1 as a contrast to S2.

S4: 'But', the sense of this sentence makes 'but' refer back to S1. S4 is actually the TS.

Paragraph 3 is an 'upside-down' paragraph, where the TS is left to the end, to lend emphasis to the main point of the whole text.

You should now be able to work out the way the writer develops his theme:

'Client is initiator of draft, so should see it' (Paragraph 1).
'Context determines how this should happen' (Paragraph 2, develops 1).
'However, not always necessary to give client a draft' (Paragraph 3, contrasts with 1)...
'though advisable because client initiates and pays for work' (Paragraph 3, returns to main theme—TS, paragraph 1).

You will also have noted that the topic sentence is not always the first sentence in the paragraph.

EXERCISE 3.3

Now we want you to try this exercise on some of your own writing.

(a) Take a piece of your writing, about three paragraphs. Give a copy to a partner. You should both try to analyse the paragraphs using tree diagrams. Do this separately.

(b) Compare your analyses. How similar are they? Where do you differ? Is your partner taking in the information you intended to convey? Always give more weight to your partner's comments than to your own. After all, you know what you intended to write. Your partner only knows what they understand from reading your work.

(c) When you have finished comparing your tree diagrams or notes, try analysing some of your partner's writing.

(d) If necessary, now rewrite your paragraphs so the meaning you intend to convey is the meaning understood by your reader.

(e) What has this exercise taught you about your written communication?

Paragraph analysis is an important exercise for looking at how we and others manage (or fail) to express often complex ideas with clarity and coherence. The writing on the page is necessarily linear, but the ideas you want to express are often in a hierarchical relationship to each other. Identifying the various levels of sentences in a paragraph illustrates this clearly. Paragraph analysis enables us to see and understand the strategies available for showing these relationships.

3.5.4.8 Long and complex sentences

What makes a 'long sentence' is quite difficult to define because readers have different levels of reading ability. While one person might find a 12-word sentence difficult to cope with, another might regard it as simplistic and patronising. The average reader seems to be happy with sentences of 15 to 20 words, but this is a rough estimate. The degree of understanding will also depend on the complexity of the sentence. By this we mean the number and length of subordinate clauses. Subordinate clauses are 'sentences within sentences' and consequently make more demands on the reader's memory.

> Read the following. How many subordinate clauses can you identify? A subordinate clause is a way of introducing additional ideas into the basic sentence using words like 'if', 'which', 'who', 'because', 'when', 'although', 'before', 'after', etc.
>
> Mr Smith, who is wheelchair-bound because he recently had an accident, is unable to come to these premises, although I have several times offered to send a taxi, which the firm is happy to pay for, and would prefer to meet us at his home if you are agreeable.
>
> *Hint*: The basic sentence is: 'Mr Smith is unable to come to these premises and would prefer to meet us at his home'.

You will see that there are seven separate ideas in the sentence, each of which could be written as a simple sentence. However, if you were to do that you would find that what you wrote appeared childish. To avoid that, the writer has chosen to use five subordinate clauses. Nevertheless, we do not consider the sentence satisfactory. It is poorly planned, as if the writer simply kept on adding bits to it as the thoughts occurred. The result is that in the middle of the sentence there is a subordinate clause 'although' which has its own subordinate clause attached to it 'which the firm is happy to pay for'. All this extra information puts an additional burden on the reader's ability to understand what is being communicated because it constantly competes with a need to understand the main message. This kind of mental gymnastics is not only demanding for the reader, but it often obscures understanding.

Take another look at the sentence. This time we have italicised the subordinate clauses.

> Mr Smith, *who is wheelchair-bound because he recently had an accident,* is unable to come to these premises, *although I have several times offered to send a taxi, which the firm is happy to pay for,* and would prefer to meet us at his home *if you are agreeable.*

What, then, are we to do? We don't want our writing to appear simplistic; nor do we want to confuse our readers by cramming too many ideas into one sentence. We suggest that the sentence could be rewritten as three sentences instead of one, as follows:

> Mr Smith is unable to come to the office because he has recently had an accident and is confined to a wheelchair. We have offered to send a taxi for him at the firm's expense, but he is still unable to come. He is happy to meet us at his home if you agree.

Note that none of these three sentences has more than three ideas in it. The real issue is not so much about how many ideas you include in one sentence, but that they should be planned and structured to help understanding, not obscure it.

3.5.4.9 Using 'correct' grammar

There is a debate in educational and professional circles about what constitutes 'correct grammar'.

See if you can write down in a few sentences what you understand by the term 'grammar'.

The word 'grammar' has many meanings, some popular, some technical. For linguists and others who understand language, the main meaning of 'grammar' is a *description* of the rules that underlie a user's ability to understand, speak and (possibly) write a given language. In its popular sense, 'grammar' is a set of rules which *prescribe* how users should speak and write. Thus we have the distinction between 'correct' and 'incorrect' English. Bear in mind that there are many varieties of English, (such as American English, Cockney, Glaswegian), each of which can be described and therefore can be said to have a grammar. However, most varieties of English are spoken varieties. The written form is expected to be usable by all speakers of English no matter which variety they learned as children. This written form of English is often referred to as 'standard English'. It is the variety of English used by all the national institutions, taught in schools, learned by non-English speakers and used in international communication.

We often import aspects of the variety we learned as children into the standard variety. Typical examples would be the restricted forms of the verb 'to be'—'they was, you was, we was'—or the double negative, 'you don't know nothing about it'. While these utterances in no way obscure the meaning of what is being conveyed, they are not regarded as acceptable in standard English, particularly in writing. Standard English has prestige because it is perceived as the language of the educated. Many people therefore believe it is superior to other varieties, so that non-standard forms are 'incorrect'. However, we prefer to think of grammatical forms as 'appropriate' or 'inappropriate', rather than 'correct' or 'incorrect'.

Moreover, since the rules of language are constantly adapting to changes in language use, what is considered 'correct' at one time may be out of date some years later. For instance, you may have been told at school that the word 'different' should be followed by 'from'. Modern usage has replaced 'from' with 'to', so'different to' could now be considered appropriate. Although it would make life a lot simpler if there were a hard-and-fast set of rules, the effect would be to freeze the 'official' language in time, and it would not take account of the process of continuous change that language is undergoing. In the end the 'rules' would simply be disregarded. Compare the fate of spelling in English! More on this later.

When a language change is identified, it is usually difficult to know when it becomes appropriate to incorporate it into the standard language. A good example is the distinction between 'less' and 'fewer', as in:

> There are less people here than there were last year.

In the traditional, prescriptive view of grammar 'fewer' is the 'correct' form. However, the use of 'less' is now so prevalent that we cannot be sure whether it may safely be used or not. David Crystal in *Rediscover Grammar* (Harlow: Longman, 1990) says that the use of 'less' instead of 'fewer' is 'widely criticised' (p. 119).

Reactions to new usage can be so extreme as to make people ill! For years now there has been debate about the word 'hopefully', as used in, 'Hopefully this case will soon be settled' instead of the traditionally acceptable 'I hope that', or 'It is hoped that'. In *Good English and the Grammarian*, Sidney Greenbaum tells of the responses of members of a panel of authors and editors to this usage:

> It is barbaric, illiterate, offensive, damnable, and inexcusable.
> The most horrible usage of our time.
> I have sworn eternal war on this bastard adverb.

'Hopefully' so used is an abomination and its adherents should be lynched.
This is one that makes me physically ill.

Fortunately, help is at hand. There are several useful reference books you can turn to for advice. The best known is H. W. Fowler, *A Dictionary of Modern English Usage*, 2nd ed. (Oxford: Oxford University Press, 1996). David Crystal's *Rediscover Grammar* (Harlow: Longman, 1990) and *Discover Grammar* (Harlow: Longman, 1996) are both up-to-date, readable grammar books. *Chambers Essential Grammar and Usage* (Edinburgh: Chambers Harrap, 1999) is more comprehensive and detailed. *An Introduction to English Grammar* by S. Greenbaum (Harlow: Longman, 1991) includes useful sections on punctuation, spelling, usage and style, together with exercises.

You will find a complete list of books at **3.11**.

How comfortable would you feel about using the following in your writing?

(a) Anyone who thinks *they* have reasonable grounds for complaint should send *their* objections to . . .

(b) If you require any further assistance, please contact *myself* or my assistant.

(c) I shall be unable to meet you on the 17th *due* to circumstances beyond my control.

(a) The traditional rule is that pronouns ('they') or determiners ('their') which refer back to a singular indefinite pronoun ('anyone') must be singular: therefore 'they' and 'their' must be replaced with 'he' and 'his'. Greenbaum considers this rule the 'traditional choice for formal writing' (p. 144).

Fowler rejects the use of the plural pronoun and determiner 'they' and 'their' because it 'sets the literary man's teeth on edge' (p. 404). He says further: 'Have the patrons of "they" etc. made up their minds yet between "Everyone was blowing their noses (or nose)" and "Everyone were blowing their noses?"'

Today, however, many writers use 'they' and 'their' because they are commonly used as singular pronoun and determiner in the spoken language. Unfortunately this usage is far from being generally accepted.

Fowler goes on to reject 'he/she' and 'his/her' which, although 'correct', are 'so clumsy as to be ridiculous'. He recommends 'he' and 'his': 'where the matter of sex is not conspicuous or important the masculine form shall be allowed to represent a person instead of a man'.

This is a convention you will be familiar with from the interpretation of statutes and other legal documents. However, as Mark Adler says in *Clarity for Lawyers* (London: Law Society 1990), men generally do not realise how irritating women find this traditional assumption that everyone is a man: 'Nor are they mollified by the answer that the Law of Property Act provides that the "masculine includes the feminine"' (p. 71). Today many writers prefer to use the gender-neutral 'they' and 'their', etc., where possible. You might choose instead to use roughly equal numbers of 'he', 'his' etc. and 'she', 'her', etc. in your writing, or alternatively to rephrase the sentence to avoid the problem altogether.

So sentence (a) above might be rephrased thus:

'If you think you have reasonable grounds for complaint you should send your objections to . . .'

(b) The reflexive pronouns ('-self', '-selves') may be used to give emphasis, as in:

Are you sure he received the letter? Yes, I took it round there myself.

The traditional rule prohibits their use as a substitute for the object pronouns 'you', 'me', 'him', 'her', 'us', 'them', though this usage is now quite common in both spoken and written English and is often found in solicitors' letters. It is a usage which makes many people feel particularly uncomfortable. We could rephrase the sentence as follows:

If you require any further assistance, please contact me or my assistant.

(c) Prescriptive grammar requires 'owing to' after a verb phrase, as in:

The match *was cancelled owing* to the bad weather, and 'due to' after a noun phrase, as in,

The cancellation of the match was due to bad weather. However, many speakers and writers now regard these usages as interchangeable. If you are in doubt about either of them you can always play safe and use 'because of'.

You are lawyers, not language experts, but the way you use language is critical to your professional success and standing. You will therefore want some certainty about what is and is not acceptable. We recommend that you select a work of reference and keep to the advice that it gives. Remember, however, that it will be updated from time to time to reflect changes in acceptable usage. You will need to update, too.

3.5.4.10 The sentence

Failure to write complete sentences is a common feature of poor writing. Here are three examples in letter-writing:

- *Thanking you* for your kind attention.
- *Assuring* you of our best attention at all times.
- *With reference* to your letter of 16th May.

These phrases fail to observe the basic rules governing the grammatical make-up of sentences. But what is a sentence? You probably think you know one when you see one, but could not provide a satisfactory definition. In fact the 'sentence' is very difficult to define. Some of you might say it begins with a capital letter and ends with a full stop. This (inadequate) definition fails to appreciate that we do not signal capital letters when we speak, and only indicate the existence of a sentence end by our intonation. Moreover, the words between the capital letter and the full stop still may not form an acceptable sentence.

Others will define a sentence by saying that it must be grammatical, so that 'The cat sat on the mat' is a sentence whereas 'Sat cat the on mat the' is not. Yet others think the sentence expresses a complete thought and so can stand on its own. This can also cause problems. For example, are these sentences? 'Yes!' Or 'Not guilty'. And are they grammatical?

The 'grammatical' definition is the least inadequate of the three we have mentioned, and therefore we will use it as our starting-point. In general, all written sentences should contain at least a subject and a verb.

The Subject

This usually appears before the main verb in statements and after the main verb in questions. It is the person or thing doing the action. For example:

The Court ordered her to pay a £50 fine.
How much did *The Court* order her to pay?

The following can be subjects:

a noun:	*Love* is a many splendoured thing.
a noun phrase:	*The cost of petrol* is going up.
a pronoun:	*It* shouldn't be allowed. *Who* killed Cock Robin?
a subordinate clause:	*What she said* was taken down in a statement.

In some formula sentences the subject is implied rather than stated, for example, [I] 'Thank you'.

Commands also omit the subject, which is usually 'you'. For example:

When you arrive, [you] go up the stairs to my office.

The Verb

The verb is a 'doing' word which describes actions and events. For example: Brian *will meet* you when you *arrive* and *take* you to court 2.

It must take an appropriate form to suit its subject. Fortunately, in English there is little variation in the form of verbs. In the present tense, only the third person singular changes, e.g., I work, she works.

In the past tense, verbs do not change form according to the person, though you need to indicate that they are past by giving them the appropriate regular or irregular form, e.g., I worked, she worked (regular), or I knew, she knew (irregular).

The exceptions to this are:

(a) The verb 'to be', which has several forms, and we are assuming you are familiar with all of them!

(b) Some modal verbs: must, can, may, might, shall, will, ought, should, would, could—which do not change at all.

Look again at the phrases we gave at the beginning of **3.5.4.10**:

- Thanking you for your kind attention.
- Assuring you of our best attention at all times.
- With reference to your letter of 16th May.

All are unacceptable sentences. It would be possible to say 'Thank you for your kind attention' (where the subject, 'I', is implied) but 'thanking' does not imply the subject 'I', and is therefore not a verb. We have seen that 'thank' is a special case and implies the subject. 'Assure' is not a special case, so a subject must be expressed. In this case the subject would be 'I' or 'we', but the word 'assuring' is not a form of the verb which can follow these subjects. In the third example there is neither subject nor verb.

It would be possible to put a comma at the end of either of these phrases and to continue with a subject and another main verb, though this is a cumbersome way of expressing yourself:

> Assuring you of our best attention at all times, I look forward to hearing from you in due course.

You probably know most of this already, even if you are not in the habit of articulating the rules. However, if you think that you need to go into more detail, consult the books listed in **3.11**.

3.5.4.11 Punctuation

Punctuation helps the reader understand written language by breaking it up into smaller units. Understanding of the spoken language is helped by pauses, intonation and emphasis on particular words and syllables. (Language experts refer to these features as 'prosody'.) This is a very subtle and sophisticated process. As we noted earlier, these features are absent from writing. Punctuation is one way of compensating for this absence. See how this sentence can have its meaning altered by punctuation:

> The judge said the accused was the most heinous villain he had ever met.
> The judge, said the accused, was the most heinous villain he had ever met.

If either of these sentences were spoken the meaning would be clear from the intonation and emphasis.

Punctuation alone is inadequate compensation for the loss of prosody (patterns of rhythm and sound). This is where intelligent planning comes in. Some words and phrases which are clear in speech may be ambiguous in writing. For example, the use of 'only' can cause serious problems. 'I only spoke to the officer' may mean:

(a) 'The officer was the only person I spoke to'; or
(b) 'I spoke to the officer but did nothing else to him' (e.g., waved to him, hit him, etc.).

Careful planning of the word order would prevent ambiguity:

> 'I spoke only to the officer' will give meaning (a). To get meaning (b) you may need to make more radical alterations:
>
> I did nothing except speak to the officer.

How well do you think you know the rules of punctuation? When would you use:

(a) a full stop;
(b) a comma;
(c) a colon;
(d) a semicolon; or
(e) an apostrophe?

Full Stops

These are used, followed by an initial capital letter, to separate sentences. For example:

> The court ordered the defendant to pay £750,000 in damages. Her solicitor advised her to lodge an appeal immediately.

Commas

Commas separate words, phrases and clauses. They usually take the place of short pauses and changes in intonation in the spoken language. Here are some examples:

(a) Lock, stock and barrel.
We met at five, went to the pub, drank a few pints and got a taxi home.

These two examples show the use of commas to separate items in a series. The last two items are usually linked by 'and', so no comma is necessary.

(b) While I was walking down the road, a dog ran into the butcher's and stole some sausages.

In this sentence, the comma separates clauses that have different subjects ('I, a dog'). There is no comma separating the next two clauses, because the subject ('dog') is the same in both. Compare:

> While I was walking down the road, a dog ran into the butcher's, and the butcher threatened it with a meat cleaver.

(c) Emma, who lives next door, is going to university next year.

Here, commas separate a clause providing additional information or explanation. If this clause is omitted, the sentence will still make sense.

(d) Compare these two sentences:

Those men, who were allegedly involved in the shooting, were charged with attempted murder.

Those men who were allegedly involved in the shooting were charged with at tempted murder.

Note the different meanings in these two examples. The first suggests a number of men, and all of them were charged. You bracket off with commas the clause which gives you additional information. If you were to remove this clause, the sentence would still make sense.

In the second sentence there are a number of men, but of those, only some were charged. In this sentence the who clause is necessary to specify who the men are, and seriously affects the meaning if omitted.

(e) This is a thoughtful, well-written, but rather dull novel.

These commas separate adjectives that describe the same noun (novel).

(f) On the other hand, the Human Rights Act will have a major impact on English law. The Human Rights Act, however, will have a major impact on English law.

In these sentences the commas separate words and phrases used to link the sentence to a preceding sentence or paragraph.

Too many commas interrupt the flow of your writing and can be an obstacle to understanding. Here is an example:

> I am writing to inform you that I have carried out a mining search on the property you are proposing to purchase, 14 Denning Walk, Wilberforce, Nottinghamshire, which is, as you know, situated in the heart of a mining area, and, I regret to report, there is evidence of serious subsidence to the neighbouring properties.

This text demonstrates that an over-zealous user of commas may be prone to extend the length of the sentence to accommodate new thoughts as they come to mind. The comma is not a substitute for thinking and planning. To make the text more readable, you would have to do more than cut out a few commas. Here is a possible alternative:

> Dear _________
>
> Purchase of 14 Denning Walk
>
> This property is in a mining area, and I have therefore carried out a mining search. Unfortunately, this has revealed serious problems of subsidence in some neighbouring properties.

Some writers mistakenly use the comma to function as a full stop. Unless a coordinating conjunction is used (such as 'and', 'but', 'or'), the sentence will be ungrammatical:

(a) The court ordered the defendant to pay £750,000 damages, her solicitor advised her to lodge an appeal immediately.

(b) The statute covers two different situations, one applies to those on income support, the other to those with an income of more than £7,500 p.a.

You can replace the commas in (a) and (b) with full stops.

(a) The court ordered the defendant to pay £750,000 damages. Her solicitor advised her to lodge an appeal immediately.

(b) The statute covers two different situations. The first deals with those on income support, the second with those who earn more than £7,500 p.a.

Alternatively you could use a stop which is not quite as full: a semicolon or colon. You can use a semicolon where you think that a clause is too closely related to what has gone before to be cut off by a full stop. For example:

> Thank you for your letter of 15 July; we will send the information you ask for in the next few days.

Semicolons are also useful to separate items in a list and are common in legal drafting where list items are often expanded upon:

> A qualifying body shall notify the Director—
>
> (a) of any undertaking given to it by or on behalf of any person as to the continued use of a term which that body considers to be unfair in contracts concluded with consumers;
>
> (b) of the outcome of any application made by it under Regulation 12, and of the terms o f any undertaking given to, or order made by, the court; (etc.)

A colon is used to precede an explanation:

> The statute covers two different situations: the first deals with those on income support, the second with those who earn more than £7,500 p.a.

It can also be used to introduce a list, for example:

> Proof in three areas is necessary to succeed in negligence: duty of care, breach of the duty and damage.

Apostrophes

You find these in all sorts of places where the rules say they should not appear at all. We use apostrophes most frequently with noun phrases to indicate possession. In spoken English, possession is indicated by adding 's' to the 'possessor'. The 's' is retained in writing, but with an apostrophe added. This is an example of redundancy, which we discussed in **3.5.4.4**. The rules are:

(a) Singular nouns and plural nouns which do not end in 's': put the apostrophe before the 's':
- John's books;
- the judge's wig;
- the men's room.

(b) Plural nouns ending in 's': put the apostrophe after the 's':
- the judges' wigs.

(c) Singular nouns ending in 's': put the apostrophe after the 's', and you can add another 's':
- Ms Jones' last case; or
- Ms Jones's last case.

We also use the apostrophe to measure a period of time:

- three weeks' holiday, five years' schooling, one month's wages, or to refer to places and premises:

We will eat at Quaglino's.
I'll meet you at the doctor's.

Furthermore, the apostrophe is frequently used to mark elision. This occurs where two words are combined to make one. For example:

He's arriving late.
Kathy's had her car stolen.
Who'd like a custard tart?
Who's next?
Don't worry.
Mustn't grumble!

The apostrophe is left out of many notices and shop signs, such as:

St Pauls, Earls Court, Menswear.

Many people think the apostrophe is old-fashioned and unnecessary. It is probably because attitudes are changing that some people are not sure how to use apostrophes and mistakenly add them to plural nouns and verb endings:

Potatoes' and fresh bean's on special offer.
Nobody love's me.

According to Crystal this usage is 'universally condemned by educated writers' (*Rediscover Grammar*, p. 111).

Another common mistake is to confuse 'whose' (the possessive pronoun) and 'who's' (the short form of 'who is' or 'who has'), as in:

Do you know whose case that is?
Do you know who's meeting us later?
Who's got the Christie file?

Use 'whose' if the word you are using cannot be expanded to 'who is' or 'who has'; if it can be expanded, then use 'who's'. The same rules apply to 'its' and 'it's'.

EXERCISE 3.4

Fill in the blanks

(a) Who_____ car is that? It_____ Mike's. It_____ for sale, actually. It_____ recently been serviced but it_____ bodywork leaves a lot to be desired.

(b) Can you see who_____ winning? No, it_____ not that clear.

(c) Have you heard who_____ given up his job? It_____ Nick. He's moving to Manchester to live with a friend who_____ mother is very ill.

How did you get on? (See the answers on page 61.) If you got any wrong, you will certainly incur the wrath of writer Lynne Truss:

'To those who care about punctuation, a sentence such as "Thank God its Friday" (without the apostrophe) rouses feelings not only of despair but of violence. The confusion of the possessive "its" (no apostrophe) with the contractive "it's" (with apostrophe) is an unequivocal signal of illiteracy and sets off a simple Pavlovian "kill" response in the average stickler. The rule is: the word "it's" (with apostrophe) stands for "it is" or "it has". If the word does not stand for "it is" or "it has" then what you require is "its". This is extremely easy to grasp.

> Getting your itses mixed up is the greatest solecism in the world of punctuation. No matter that you have a PhD and have read all of Henry James twice. If you still persist in writing, "Good food at it's best", you deserve to be struck by lightning, hacked up on the spot and buried in an unmarked grave.'
>
> (Truss, L., *Eats, Shoots and Leaves: The Zero Tolerance Approach to Punctuation*, Profile Books Ltd, p. 43).

We wouldn't go quite that far! Nevertheless, the rules for using the apostrophe are straightforward, yet it is probably the most misused of punctuation marks. Our advice is that you should learn to use it correctly if you want to be considered an educated writer.

3.5.4.12 Spelling

Mark Twain remarked that he could not respect a man who could only spell words one way. English spelling is difficult because words are not always spelt as they are pronounced. There are two main reasons for this. One is that the spelling system introduced by the Normans was mixed with the system we used before the Norman Conquest. This accounts for two spellings of the same sound, for example, -se and -ce ('mouse' and 'mice') and two sounds for the same spelling, for example, 'g' as in 'get' and 'gem'.

The other reason is that spellings have generally not changed, while pronunciation has. During the Middle Ages the few people who could write spelt words in different ways. A fixed and uniform spelling system only began to be established in the 15th Century with the introduction of printing. The standard system developed by printers was adopted in the 18th Century by the most important dictionaries. This is largely the system we have today, unaffected by the significant changes in pronunciation which have taken place since the 15th Century.

The effect of this fossilisation is that for instance, we preserve letters in words that are no longer pronounced:

ni*ght*, *k*now, de*b*t, sa*l*mon.

We use different spellings for the same sound, such as 'ee' and 'ea' in 'meet' and 'meat', because in the past 'ee' and 'ea' were different sounds. If you compare 'greet' and 'great' you will see that we cannot count on this 'rule'. Similarly, when you consult books on spelling you will find that the 'exceptions' often outnumber the 'rules'.

Some people see poor spelling as a sign of intellectual incompetence. Since there are so few rules, however, learning to spell involves memorising rather than exercising your intellectual skills of judgement, discernment and reflection. These are the skills you need to learn the complex linguistic skills of grammar and punctuation, where even a small error can cause misunderstanding. Poor spelling is rarely responsible for a breakdown in communication.

Nevertheless, you have to recognise that poor spelling may make your reader think you are uneducated, incompetent or unprofessional. Be sure to use a dictionary or spell checker.

Reading about paragraphs, sentences, words, punctuation and spelling is not everyone's idea of a good time. As a lawyer, however, communicating clearly with lay people and other professionals is part of your business. To do it well is part of the process of providing a caring and efficient service to your clients and colleagues. To do it poorly is a recipe for confusion, disappointment, frustration and lost business. In this section we have tried to give you some important insights into the processes of effective writing, so that your written communication is clear, concise and correct.

3.5.5 Revising: the importance of self-editing your writing

We can summarise this section on strategies with a list of questions you should answer honestly every time you have composed a written document. Since this is a method of looking at your writing through the eyes of the reader, you may find it useful for a colleague to read it and answer the questions too. You can then check your self assessment with their assessment.

3.5.5.1 Self-editing checklist

Purpose:

(a) What is the purpose of the communication?

(b) Who is/are my intended reader(s)?

(c) Have I adapted style and content to suit the reader's needs?

(d) Have I dealt with the issues?

(e) Have I answered all the questions?

(f) Have I answered them in enough/too much depth?

Content:

(a) Is the information accurate?

(b) Is it relevant?

Humanity:

(a) Will my tone produce the desired response?

(b) Is it friendly, courteous, helpful, frank, or is it curt, distant, patronising, vague?

Layout:

(a) Is the layout appropriate for the purpose and content?

(b) Is it set out in manageable blocks?

Structure:

(a) Do I get to the point quickly and make it clearly?

(b) Are the sentences short enough?

(c) Does the order of sentences and paragraphs make sense?

(d) Does each paragraph contain just one main idea?

(e) Is there a link between each paragraph and the next?

(f) Are there links between sentences in each paragraph?

Language:

(a) Have I used plain language i.e., clear, concise, correct language that can be easily understood by the reader?

(b) Have I omitted words and phrases which are:

(i) infrequently used;

(ii) inelegant;

(iii) redundant;

(iv) unnecessarily technical;

(v) verbose; or

(vi) vague.

(c) Are punctuation and spelling correct?

3.6 The conventions of letter writing: an exercise

Here is a solicitor's letter which contains many of the characteristics of poor writing we have discussed. Note down in detail what you think is wrong with it. If necessary, use the self-editing checklist to help you. Then compare your comments with our 'annotated' version. Finally, rewrite the letter clearly, concisely and correctly.

> Dear Both,
>
> Re: 14 Denning Walk, Bingham, Oxfordshire OX13 4QC
> Sao Miguel, Bridge St, London SW13 9JJ
>
> Further to previous correspondence on this matter, we now enclose herewith the enquiries and fixtures and fittings list, to be duly completed by yourselves and returned, forthwith, to this office.
>
> As regards title to the property. It is registered at the Land Registry with absolute freehold title. You will observe from the Land Registry plan that there is a small area of land, coloured pink, and within that land there is a Water Main and I am enclosing a copy of a Deed of Grant dated 10th May 1966 which gave the right to the then Water Board to lay and maintain a water pipeline through that land. Therefore the Water Company have a legal easement, which means the main can remain there.
>
> It follows from that, if the Water Main ever needs to be repaired then the Water Company has the right to enter upon the land in order to maintain and repair the said pipe.
>
> They would have to cause as little damage as possible and put right any damage caused afterwards, in so far as such is possible. It also follows that you would not be able to build over the Water Main. There are no Covenants which affect this property other than those contained in the aforementioned Deed of Grant dated 10th May 1966 related to the Water Main.
>
> As far as exchange of contracts prior to completion is concerned, you will be required to pay a deposit to the Vendors Solicitors. As yet, we are not entirely clear how much Deposit will be received on the Sale, but should be able to provide the same amount on your purchase of this property. However, it is not inconcievable that, the same sum will not be able to be provided, or that the amount will be insufficient. Should this be the case, it will be brought to your attention with all speed and alternative arrangements will be put into effect.
>
> You should contact this office if there are any difficulties arising from this letter.
>
> This is to keep you informed.
>
> Yours faithfully,
>
> ____________

Many of the problems with the letter arise from the failure to consider the needs of the reader. The sentences are full of verbose, redundant and technical words and phrases. Poor structure and layout reflect the writer's failure to think and plan.

Furthermore, it is not clear how formal or informal this relationship is supposed to be. We know there is a relationship between lawyer and client already established through previous correspondence, if not through personal contact. The use of 'Dear Both' suggests that the relationship is fairly informal, whereas 'Yours faithfully' at the close of

Dear Both,

Re: 14 Denning Walk, Bingham, Oxfordshire OX13 4QC
Sao Miguel, Bridge St, London SW13 9JJ

Further to previous correspondence on this matter, we now enclose herewith the enquiries and fixtures and fittings list, to be duly completed by yourselves and returned, forthwith, to this office.

As regards title to the property. It is registered at the Land Registry with absolute freehold title. You will observe from the Land Registry plan that there is a small area of land, coloured pink, and within that land there is a Water Main and I am enclosing a copy of a Deed of Grant dated 10th May 1966 which gave the right to the then Water Board to lay and maintain a water pipeline through that land. Therefore the Water Company have a legal easement, which means the main can remain there.

It follows from that, if the Water Main ever needs to be repaired then the Water Company has the right to enter upon the land in order to maintain and repair the said pipe. They would have to cause as little damage as possible and put right any damage caused afterwards, in so far as such is possible. It also follows that you would not be able to build over the Water Main. There are no Covenants which affect this property other than those contained in the aforementioned Deed of Grant dated 10th May 1966 related to the Water Main.

As far as exchange of contracts prior to completion is concerned, you will be required to pay a deposit to the Vendors Solicitors. As yet, we are not entirely clear how much Deposit will be received on the Sale, but should be able to provide the same amount on your purchase of this property. However, it is not inconcievable that the same sum will not be able to be provided, or that the amount will be insufficient. Should this be the case, it will be brought to your attention with all speed and alternative arrangements will be put into effect.

You should contact this office if there are any difficulties arising from this letter.

This is to keep you informed.

Yours faithfully,

____

the letter is a formal closure which normally follows the opening 'Dear Sir' or 'Madam'. Twice in the letter the client is asked to contact the office rather than the writer personally or another named person. This, together with the use of the passive, makes the tone impersonal and peremptory.

The writer is inconsistent in other ways. In the first paragraph the writer is 'we' but 'I' in the second. Again, the readers may be confused about the kind of relationship they have with their solicitor, and who is actually doing the work: the solicitor or some other person in the firm? You can clarify this by using 'I' if you want to encourage a fairly informal, friendly relationship, and 'we' when writing about the general undertakings of the firm, e.g., 'We aim to provide a quick and efficient service'.

To be consistent you should then sign personally at the bottom. You can sign in the firm's name if you want to keep the recipient at a distance, though you should give the name of someone in the firm to contact.

Like newspapers and magazines, firms of solicitors have their own 'house style'. Your firm's style may therefore determine how you sign letters and the level of formality you adopt.

The letter writer is also inconsistent in the use of capital letters. These are not confined to names and titles but are sprinkled liberally throughout the text for some unknown purpose—perhaps to venerate some noble legal institutions: 'Solicitor', 'Deed of Grant', 'Vendor', 'Sale'? This cannot account for 'Water Main', however.

Tinkering with this writer's prose will improve it, but not much. The whole text needs radically rethinking. You could probably get the message across more clearly by dealing with sale and purchase in separate letters.

Compare our rewritten version with yours:

Dear Mr and Mrs ______,
Sale of 14 Denning Walk
Purchase of Sao Miguel

Before we can agree a contract of sale with the purchasers, we have to give them some more information. I therefore enclose two forms for you to complete.

The Enquiries Before Contract form provides the purchasers with important details about the property. The fixtures and fittings list tells them which items you are taking with you and which you are leaving behind. Could you therefore complete both forms and return them to me as soon as possible in the envelope provided.

I also enclose a Land Registry plan of Sao Miguel. The house is registered with absolute freehold title. This means that you will be the owners of the house from the date of completion.

You will notice on the plan that a small area of the property is coloured pink. This marks a water main running through the property. You may not build anything over this pipe. Moreover, the water company has the legal right to enter the property to repair and maintain the pipe. This right is given by a deed of grant dated 10th May 1966. The deed also protects you by insisting that any work the water company carries out causes as little damage and inconvenience as possible. The company must put right any damage it does cause to the main. This is the only right over the property granted to someone other than the owner.

When I exchange contracts on your behalf I have to pay a deposit to your vendor. I expect to use the deposit I receive from your purchaser to do this. If your vendor considers this amount too small, you may have to arrange a loan for the shortfall. This is normally not a problem and I will be happy to advise you.

Please telephone me if there are any points in this letter you wish me to clarify.

Yours sincerely,

____

3.6.1 Common pitfalls in letter writing

As well as the emphasis we have put on clear, concise and correct writing and the importance of thinking and planning, there are some points specific to letter writing that many people have difficulty with:

The heading

Keep this short and descriptive. A good heading should indicate the main point of your letter. Using the word 're' ('in the matter of') in headings is not good style. Not only is it redundant, it's not even English!

Opening sentences and paragraphs

Frequently letters begin with inelegant, redundant phrases. Often these are not full sentences. We believe writers do this because they think it is necessary to use a polite and courteous opening, rather than getting straight to the point. For example:

> Further to our telephone conversation yesterday, in which you requested information regarding the sale of your business, we are now in a position to inform you...
>
> With reference to your letter of 13 December regarding your claim. It should be pointed out that...

Whilst we would agree that courtesy is crucial, it can be expressed more elegantly and concisely. For example:

> I am now able to provide the information you asked for in our telephone conversation yesterday.
>
> Thank you for your letter of 13 December. I would like to draw your attention to...

Our opening sentence in the sale and purchase letter above goes straight into the information to be conveyed without referring to previous correspondence between lawyer and clients. You may think it sounds a little abrupt. We thought that in a conveyancing relationship like this, where information is constantly passing between them, such an opening was unnecessary.

The principle is to keep your opening short and simple, get to the point of the letter quickly and avoid formulaic phrases (especially those that are not sentences!).

Closing sentences and paragraphs

The same principle applies here. For example:

> Please contact me if you have any questions.
> I look forward to meeting you on (date).
> If you need further information, I shall be happy to supply it.

In summary:

Avoid	Use instead
Please find attached Please find enclosed Below please find	You will see that I have attached enclosed Here is a copy of
In respect of Regarding With regard to	About Change the sentence structure to avoid!
Further to With reference to Referring to	Thank you for I am writing to confirm/provide you with I refer to (if you must)
Re	Nothing
Assuring you of our best attention at all times Thanking you for your kind attention Thanking you in anticipation Looking forward to meeting you (Note that none of these are sentences)	Nothing Nothing Nothing I look forward to meeting you
If you require further information please contact myself or my assistant	Please contact me if you would like any further information

3.7 Writing e-mails

Almost all businesses and other organisations use e-mail nowadays. You will therefore probably find that you will need to use it more and more as your career progresses. It is mainly used among colleagues at present, but the time may not be far away when you regularly use it to contact clients.

Writing e-mails is subject to a few basic rules:

(a) Take the same amount of care writing e-mails as you would writing a letter. Use the Plan, Write, Revise approach. Remember, your electronic communication is going to be stored somewhere and can be brought out later if needed. Just as with writing letters, you are communicating not only at a distance, but over time too.

(b) Know where you are in cyberspace. Some of your e-mails will be to colleagues and friends. In such cases, greater informality is acceptable, but make sure that the level of formality/informality between you and your correspondents is agreed. It is very easy to forward e-mail at the push of a mouse button, so be confident that anything you say, especially about someone else, is not going to embarrass them or you if it comes out. Electronic communication has the appearance of informality and detachment because we are interacting with a computer monitor. It is therefore easy to forget that another human being is going to read what we write. Moreover, once the e-mail is sent, we have no control over who else receives it.

(c) Say what you need to say in one go. Because e-mail is so immediate, it is easy to fall into the habit of sending off a communication the moment you think of something to say. Try to avoid this. Compose your e-mail as carefully as you would compose a letter and send all the information you want to convey in one go. It is a poor expenditure of your time, and a real irritant for the recipient, to keep getting information in dribs and drabs. Don't let the ease and convenience of e-mail seduce you into being less than professional.

(d) Become familiar with the electronic security system. There are various security features present on most e-mail systems. Learn about those on the system you are using. It may well be that your firm has a policy about what can and can't be sent by e-mail. If that is the case, respect it. Make sure you are properly trained to use the system effectively and ethically.

(e) Use anti-virus software as a matter of course. Viruses and other forms of electronic contamination are rife. Most of us have unwittingly downloaded viruses, usually through our e-mail programmes. The main thing to watch out for are attachments to e-mails. The sensible advice is never open an attachment to an e-mail, even if it is from someone you know and trust. The 'Love-bug' virus sent out in May 2000 got into recipients' address books and sent itself as an attachment to all the names it found. Consequently, many recipients opened the attachment, believing to come from a known and trusted source. What you should do is save the attachment to a removable disk and scan it for viruses. When you are sure it is a genuine attachment, then you can open it.

(f) Don't use cyber-abbreviations. The development of mobile phones which can take text messages has led to the use of abbreviations to save time and space on the phone display (e.g., IMO, meaning in my opinion; AFAIK, meaning as far as I know; :) meaning ☺; etc.). Keep these for phone messages only. Don't be tempted to transport them onto desktops or laptops.

(g) The use of uppercase only in e-mail is the equivalent of shouting! DON'T DO IT!

3.8 Testing your writing ability

In this chapter we have split 'Writing' up into a number of discrete components. However, effective writing is really about integrating all of them and being self-critical. You should measure the success of your writing by the ease with which your reader understands the information you give them, and whether things you want to happen as a result of your writing actually do happen, when you want and in the way that you want.

We therefore conclude with a redrafting exercise that combines a number of the components we have discussed.

EXERCISE 3.5

Work alone or with a partner:

(1) Identify what is wrong with the sentences below.

(2) If possible, give reasons for your answers.

(3) Rewrite each sentence so that it is clear, concise and correct.

(a) I hope that the above satisfies your requirements. If you have any questions, please do not hesitate to contact myself or my secretary.

(b) The defendents principle argument was that he did indeed fulfill each and every criteria set out in the regulations. But this definately failed to effect the decision of the court.

(c) Our advise is based on the following facts, as described by you; whether you are, by law, allowed to disclose information regarding the violation of health and safety regulations, with respect to the display frequency of computer screens produced by your employer, and what will be the consequences of your disclosure, and, in particular, whether you will have the right of reinstatement, in the event of the termination of your contract, due to the disclosure.

(d) Regarding our conversation in my office last week, below please find my comments relating to the issues, which we discussed during our meeting.

(e) The court then turned to the request which had been made by the Claimant for the issue of an injunction. With regard to that request, the argument was made by the defendant that the injunction relief was not necessary because of the fact that the exclusion clause was already null and void by reason of the prior order of the court. That being the case, the exclusion clause could have no further force or effect and the defendant argued that in such an instant full and complete relief could be given without the issue of an injunction. The court found itself in agreement with that argument.

(f) My client is willing to settle this claim for £15,000, to be paid by your client and your client must immediately return the plans and specifications and must remove all of It's equipment from the premises. Further, my client insists upon having replaced the entire section of fence which your client took down, the replacement to be at your clients expense.

(g) I have a list of the shareholders of the Company, who's participation in the share capital exeeds 10%

(h) With reference to your letter of 14 February. I would like to confirm that the best manner of protection of your interest against non-licenced producers' of the software is an injuction, being a legal instrument commonly used in such cases.

(i) (A radio news item) The body of a man was found by a burnt-out car.

(j) This agreement may be terminated by either party, dependant on thirty days notice being given to the other party.

(k) Prior to the collection of your vehicle, please insure that you pay for your parking at the automatic machines located at the administration building which is situated in the vicinity of the exit.

(l) I have given implicit instructions to my staff to keep noise to an absolute minimum due to the close proximity of residential properties.

(m) In accordance with your instructions we have now prepared a draft agreement between your Company and Mercury Promotions Limited and we enclose the same herewith for your perusal and consideration.

(n) We would refer you inter alia to paragraph 4 of the said agreement and ask you to note that we have provided that re-instatement is to be affected at the expense of the advertiser. We ask you to confirm that such clause is in accordance with your instructions.

(o) Information contained in this form and on the passport record to which this application form relates may be passed to other government organisations and law enforcement agencies for the purpose of checking your application and in the subsequent use of any passport issued as a result of this application.

(p) Please note that if during the validity of a passport on which a child is included a new passport is required, for reasons of loss or amendment, a separate passport for the child would need to be issued.

3.9 Learning outcomes

In this chapter we looked at:

- the 'Plan, Write, Revise' approach to effective writing;
- strategies for improving your writing;
- how to vary your language and style to suit the needs of your reader;

- being self-critical and continuing to develop your writing skills.

You should now be able to:

- list the strengths and weaknesses of your own writing;
- identify and put into practice strategies for improving your writing;
- plan your writing to take account of the needs of your reader;
- vary your language and style to suit the needs of the content and your reader;
- use reference books and other aids for effective writing;
- write so as to avoid using discriminatory language;
- prepare and draft a range of documents to a variety of readers.

3.9.1 Law Society written standards

Throughout your career as a solicitor you will be constantly reviewing and improving your skills. The Law Society's standards for writing are combined with those for drafting. We have reproduced these in full below.

3.9.1.1 Writing and drafting

Students should have developed a basic skill in the preparation and drafting of a range of documents and should be able to formulate and present a coherent piece of writing based upon facts, general principles and legal authority, in a structured, concise and where appropriate, persuasive manner.
Students should be able to draft documents that:

(1) meet the client's goals, carry out the client's instructions and address the client's concerns;
(2) maintain a standard of care which protects the interests of the client;
(3) deal appropriately with client care and professional conduct issues;
(4) accurately address all the relevant legal and factual issues;
(5) where appropriate, identify relevant options, including the costs, benefits and risks of those options;
(6) where appropriate, demonstrate a critical use of precedents;
(7) are logically organised;
(8) form a consistent and coherent whole;
(9) follow the rules of grammar;
(10) demonstrate appropriate use of language;
(11) are succinct and precise;
(12) meet any formal legal or other requirements.

Students should be able to explain the principles and criteria of good drafting practice and should be able to:

(1) identify pieces of drafting where those criteria are either met or not met;
(2) modify their own drafting errors to meet these criteria;
(3) transfer drafting principles learned in one context to other contexts;
(4) use techniques for appraising and developing their own writing and drafting style.

Students should have developed a basic skill in the use of precedents, in setting out documents, in drafting clauses, and should be able to write letters to lawyers, other professionals, clients and other non-lawyers.

3.10 Self-test questions

(1) What are the key differences between the spoken and written language?

(2) What are the implications of this for effective writing?

(3) How do you ensure that your writing is appropriate for your audience?

(4) Outline the structure of a typical paragraph.

(5) Draw up a guide to plain English writing. Use the list on page 33 as your starting point. Compare your guide with a partner's. Are there any similarities? Are there any differences?

(6) Consider the following two forms of written communication:
(a) an academic essay; (b) a letter to a client.
What principles of good writing are common to both?
What differences in approach would you need to make for each?

(7) List (a) the strengths and (b) the weaknesses of your own writing. What changes will you make?

3.11 FURTHER READING

Our aim in writing this chapter has not been to provide you with a comprehensive course in legal writing. Instead we have tried to draw your attention to some of the features of poor writing that have frustrated, infuriated or confused us as readers. Writing well is a skill worth developing for both professional and personal reasons. If you want to commit yourself to further development, here is a list of useful (and, we hope, readable) titles to help you.

Asprey, M., *Plain Language For Lawyers* (Sydney: Federation Press, 1996).

Ayto, J., *The Oxford Essential Guide to the English Language* (Oxford: Oxford University Press, 1998).

Burchfield, R. W., *The New Fowler's Modern English Usage* (Oxford: Oxford University Press, 1996)

Chambers Essential Grammar and Usage (Edinburgh: Chambers Harrap, 1999).

Crystal, D., *Rediscover Grammar* (Harlow: Longman, 1990).

Crystal, D., *Discover Grammar* (Harlow: Longman, 1996).

Cutts, M., *Plain English Guide* (Oxford: Oxford University Press, 1999).

Fowler, H. W., *Modern English Usage*, 2nd ed. revised by Sir Ernest Gowers (Oxford: Oxford University Press, 1996).

Garner, B. A., *The Redbook: A Manual of Legal Style* (St Paul, MN: West Group, 2002).

Gowers, Sir Ernest, *The Complete Plain Words* revised by Greenbaum and Whitcut (London: HMSO, 1986).

Greenbaum, S., *An Introduction to English Grammar* (Harlow: Longman, 1991).

Mellinkoff, D., *Language of the Law* (London: Little, Brown, 1963).

Rylance, P., *Legal Writing and Drafting* (London: Blackstone Press, 1994).

The National Consumer Council, Plain English Campaign and Clarity (a movement for the simplification of legal English) have produced guidelines on plain English for letter writing and drafting. Useful publications are:

Adler, M., *Clarity for Lawyers* (London: Law Society, 1990).

National Consumer Council, *Plain English for Lawyers* (London: National Consumer Council, 1984).

Plain English Campaign, *Language on Trial: The Plain English Guide to Legal Writing* (London: Robson Books Ltd, 1996)

Wydick, R., *Plain English for Lawyers* (Carolina Academic Press, 1985).

ANSWERS TO EXERCISE 3.4

(a) Whose car is that? It's (it is) Mike's. It's (it is) for sale, actually. It's (it has) recently been serviced but its bodywork leaves a lot to be desired.

(b) Can you see who's (who is) winning? No, it's (it is) not that clear.

(c) Have you heard who's (who has) given up his job? It's (it is) Nick. He's moving to Manchester to live with a friend whose mother is very ill.

4

Drafting legal documents

4.1 Learning outcomes

This chapter deals with the skills needed for drafting legal documents. In this chapter we focus particularly on:

- understanding and being able to plan your preparation for drafting;
- understanding and being able to consider, select and use precedents;
- understanding and being able to undertake the drafting and amendment process.

4.2 Introduction to drafting

A popular joke about lawyers asks, 'What do you get if you cross the Godfather with a lawyer?' Such is the confusion often sown into the minds of lay clients by a consideration of legal documents that the response to the question should come as no surprise: 'A man who makes you an offer you can't understand.'

Legal drafting has often been the source of amusement. Take, for example, the central plot of the Gilbert and Sullivan opera, *The Pirates of Penzance* in which a youth, Frederick, is indentured to be an apprentice to the Pirate King. Frederick interprets the document as releasing him from his bounden duty upon reaching his twenty-first year, whereas the document states the release to occur on his twenty-first birthday—a grave concern as Frederick's birthday was on 29 February and thus he only has a birthday every leap year. 'A paradox, a paradox, A most ingenious paradox!' and a clear illustration of the ramifications of infelicitous and thoughtless drafting.

The typical product of drafting is a document embodying an agreement or accord between parties, often separately advised and represented, and intended to regulate the legal relationship between those parties. A well-drafted document should marry together the relevant facts and law in a clear and concise manner and encapsulate the final agreed instructions or negotiations.

All too often documents, like the indenture from *The Pirates of Penzance*, fail in their purpose, because of incorrect wording. Sometimes an eventuality which should have been considered has been overlooked. Sometimes the document is so obscurely drawn that the court must be asked to construe it.

EXERCISE 4.1

What is the difference between a drafted legal document and a letter or memorandum?
Consider a standard contract such as a credit card agreement or insurance policy and look at the structure of the draft. What do you notice?

4.3 The written standards

The Law Society's written standards for the Legal Practice Course require that students should be able to prepare a range of documents and formulate and present a coherent piece of writing based upon facts, general principles and legal authority in a structured, concise and, when appropriate, persuasive manner. Students should be able to draft documents that:

(1) meet the client's goals, carry out the client's instructions and address the client's concerns;

(2) maintain a standard of care which protects the interests of the client;

(3) deal appropriately with client care and professional conduct issues;

(4) accurately address all the relevant legal and factual issues;

(5) where appropriate, identify relevant options, including the costs, benefits and risks of those options;

(6) where appropriate, demonstrate a critical use of precedents;

(7) are logically organised;

(8) form a consistent and coherent whole;

(9) follow the rules of grammar;

(10) demonstrate appropriate use of language;

(11) are succinct and precise;

(12) meet any formal legal or other requirements.

This chapter will introduce you to the relevant skills and matters relating to drafting by considering the various stages in drafting a document:

(a) Preparing to draft;

(b) Responsibility for drafting;

(c) Getting down to drafting;

(d) Appearance, style and content of the draft;

(e) Your draft in their hands;

(f) Use of grammar and language;

(g) The process of amendment;

(h) Engrossment and completion;

(i) Construction of documents;

(j) Plan, write, revise;

(k) Persuasive and informative drafting.

4.4 Preparing to draft

4.4.1 Importance of preparation

As with any other lawyerly skill, preparation is essential to successful drafting. Unlike advocacy or negotiation, it is true that you will have the option to reconsider your document before presenting it for public consumption. However, it is certainly true that without adequate preparation the document will take longer to draft and is likely to omit important matters.

4.4.2 Identify the client's goals, concerns, instructions

All documents produced by a lawyer are produced in order to carry out a client's instructions. The lawyer must work out with the client what needs to be done, applying legal knowledge and skills to the client's business or personal situation.

In a commercial matter, the solicitor should be familiar with the whole range and objectives of the client's business so as to understand precisely the agreement which has to be drafted. Preparing a will entails the solicitor obtaining background information about family, dependants and property for which the client may see no need, but which the experienced solicitor recognises to be necessary to attain the client's objectives.

As with any work, it is important to check that instructions have been given by the appropriate person. Does an employee have the authority to instruct you? Is the person instructing you to deal with property the legal owner of the property? If not, make sure the instructions are checked with the appropriate person or people.

It is also always necessary to consider whether the instructions can be accepted at all. If you were to prepare the document, would there be a conflict with an existing client's interest? Is the document you are being instructed to prepare illegal? You must of course consider whether you can accept the instructions in light of professional conduct rules.

When taking instructions, do not assume that the person instructing you knows the law. Your job is to arrive at the correct legal solution to the client's problem. You may appreciate that there may be more than one solution, and you must identify with the client the client's primary goals and concerns. The client may approach you wanting an agreement to buy Blackacre completing in one year's time when the client can obtain planning consent. This can be achieved by entering into an option agreement, a conditional contract or an unconditional contract, each with its own advantages and disadvantages depending on your analysis of the client's goals and concerns.

Before drafting it may be desirable to make a site visit. Especially in relation to complex commercial property matters it is of enormous assistance to do this so that you know the layout of the land when drafting a lease or agreement. It may also be beneficial to the solicitor drafting commercial agreements to visit the client's business to form a clearer view of it. The cost of making a site visit should be agreed with the client beforehand.

4.4.3 Analyse all the relevant legal and factual issues

You must assimilate and analyse the actual factual context according to your instructions. Only then can you formulate a clear idea of what the draft document should contain.

Before setting pen to paper (or more likely settling down with keyboard and mouse in hand), consider your instructions. Are they clear to you? What is the objective of the draft? If it is a contract, can you identify the parties, the consideration, the obligations of each party, any conditions, warranties and representations, any provisions which take effect after a breach of contract and any other material matters on which you have instructions? If there are gaps, ask yourself why there are gaps. Will they be filled by other documents? Are they matters which neither you nor your client have considered? Do you need further instructions?

If you conclude that you need further instructions, consider whether they are necessary before you can prepare the first draft or whether you can produce a draft and afterwards ask your client to fill in the missing piece.

Equal importance should be afforded to consideration of the law affecting the agreement to be prepared. Are there any restrictions to what is being proposed? Suppose you are drafting a director's service agreement. Is it lawful to bar the director from working anywhere in the UK for five years after dismissal? Check up by researching the texts and original sources.

Has the law altered recently? Using precedents as a base is no safeguard if they are out of date. Taxation is an obvious area where the law is always changing. Is what is proposed tax-efficient or should the client's affairs be ordered in a different manner?

EXERCISE 4.2

Go to the *Encyclopaedia of Forms and Precedents*. How would you know the date on which any precedent was prepared? How would you update your legal knowledge? Is the precedent more up to date in its online version?

You must be aware of any provisions which imply obligations into certain types of contract and whether you can override them. For example, when drafting a partnership agreement, the Partnership Act 1890 must be considered. Section 24 provides that the interests of partners in partnership property and their rights and duties in relation to the partnership are to be determined by the rules contained in that section unless there is a special agreement between the partners. Thus, if the partnership profits are not to be shared equally, your partnership agreement should specify the contrary agreement.

Another example is that there is no point in drafting a bill of lading without considering the Carriage of Goods by Sea Act 1971, which provides that the Hague–Visby Rules shall automatically apply to certain bills of lading. If the rules apply to the bill of lading, there are minimum duties and liabilities which cannot be excluded or reduced by agreement. It is otiose to attempt to prepare a bill of lading for a client which would be void.

Time spent before drafting on analysing the factual situation and the legal context into which the document is to be set is never time wasted.

Summary

(a) What are the facts, which provide the bones of the draft? Are you sure your instructions are clear?

(b) What is the law that is applicable? You must research the up-to-date law.

(c) Carry out all further research into any aspect which may affect the draft.

4.4.4 Where appropriate, identify any options

Preparatory analysis of the law and the facts may well reveal that there are alternative ways of dealing with the matter. The primary goal must be to achieve what the client wishes in a legal context. However, if this can be achieved in various ways, consider adopting an option which saves on stamp duty, or one which may be tax-efficient, or one which saves costs.

4.4.5 Precedents

Precedents can be very helpful in drafting, but they must be used with care.

A precedent should be seen as a checklist against which you ensure your draft does not omit material matters.

A precedent is not a replacement for a solicitor. It should not be merely copied without thought.

4.4.5.1 What precedents are available?

There are many excellent books and on-line packages of precedents, access to which is vital to most solicitors. Chief amongst these in the non-contentious sphere is the *Encyclopaedia of Forms and Precedents*, which is currently in its 5th edition. The *Encyclopaedia* is available as a hard copy set or online via the Lexis-Nexis Butterworths platform. Almost all firms of solicitors will have access to this series. It is ordered alphabetically and covers topics from agency to landlord and tenant, from copyright to shipping, from acknowledgements and receipts to wills and administration. There is an updating service. In addition to precedents, for each subject area, the *Encyclopaedia* has a concise statement of the relevant law and most precedents have comprehensive footnotes citing law, cases and common variant clauses. The *Encyclopaedia* includes many codes of practice and rules which are of assistance to the practitioner. For example, vol. 1 contains the The British Code of Advertising, Sales Promotion and Direct Marketing, vol. 3 contains several commonly used arbitration rules, such as the UNCITRAL rules and the CEDR model ADR contract clauses.

The aim of the *Encyclopaedia* is stated by the editor-in-chief, Sir Peter Millett (now Lord Millett), as meeting the needs of one who must 'carry into effect his client's often half-formulated intentions, . . . and do so in language which is not only legally effective and unambiguous, but also simple and concise'.

Other excellent precedents may be found inter alia in:

(a) *Practical Conveyancing Precedents* and *Practical Lease Precedents*, both by Trevor Aldridge;

(b) *Precedents for the Conveyancer*, which is a loose-leaf two-volume work, which, while primarily dealing with a conveyancer's requirements, also has sections on business transactions and charities amongst others;

(c) *Parker's Modern Conveyancing Precedents*, 2nd ed. by Eric Taylor, which in its 1st edition pioneered the use of plain English in drafting conveyancing documents;

(d) *Practical Commercial Precedents*, which is in loose-leaf format and in particular contains precedents for City-type agreements of a specialised nature;

(e) *Kelly's Draftsman*;

(f) the *Will Draftsman's Handbook* and *Practical Will Precedents*, which specialise in probate and will drafting.

In the contentious field, the primary source of precedent material for statements of case and prescribed court forms is *Atkin's Court Forms*, which is now in its 2nd edition. *Atkin's* is also carried on the Lexis-Nexis Butterworths platform. Butterworths also produce the Civil Courts Precedents.

In addition to specific precedent books, many texts and journals may have directly linked precedents. Knight on *Acquisition of Private Companies and Business Assets* has, for example, drafts of a share purchase agreement. *The Conveyancer* often has conveyancing-related precedents.

Almost every firm will also keep a know-how database, a more or less formal system of precedents developed over the years. Be familiar with the precedents available in your firm, especially those such as leases and commercial agreements which will be frequently required in practice. Familiarity will allow you to adapt these documents to suit the client's purposes.

4.4.5.2 Using a precedent as a base for drafting

Before you can find the precedent or precedent clauses most appropriate to your needs, you must have analysed the precise factual and legal context. Then read the precedent. Never simply reproduce it.

If the precedent you are working from is online or in your firm's know-how system, it is always good practice to print off a full copy. Do not try to read it online. Check whether the precedent has been corrupted with clauses altered, added or deleted either deliberately or erroneously. Research shows that people read online forms by scanning for major points rather than reading the text closely. Fonts used typically are more difficult to read online.

If the precedent you are working from is contained in a book, you should photocopy the clauses. Some writers suggest that you should rewrite and seek to rephrase a precedent. This forces you to think about the wording. However, a clause or part clause may be accidentally omitted when working this way and cannot be easily picked up once the draft has been word-processed. Working from a photocopy ensures that you always have the full draft of the basic precedent available. Secondly, it will generally be easier to read the printed text than your handwriting. This is so even if you will be typing the draft up on your PC or if your secretary will be typing it up. Thirdly, time and thus expense to the client may be saved.

EXERCISE 4.3

Use the *Encyclopaedia of Forms and Precedents*. Consider which precedent or precedents would be most appropriate to use when preparing the indenture of apprenticeship between the Pirate King and Frederick. Are there any clauses which may help you from any other precedents?

Go first to the general index. What keywords might you search? Would you suspect that 'Pirate' may not be appropriate?

EXERCISE 4.4

Consider precedent 16 on p. 414 of vol. 4 of the *Encyclopaedia of Forms and Precedents*, 4th ed. and compare it with precedent 10 on p. 91 of vol. 7 of the 5th ed. as originally published.

Both are declarations of trust of a clergy rest home. Note the modernisation of style and form in the 5th ed. precedent; but also note the continuing use of words from the 4th ed. in the 5th, e.g., 'the Trustees shall stand possessed of (trust monies) upon trust to apply the same'. There has been little attempt to rewrite in plain English words which have legal effect.

Are there any substantive differences between the precedents? Is one or other drafter relying on the Trustee Act 1925 or other legislation?

4.5 Responsibility for drafting

When drafting an agreement between two or more parties it is necessary to establish which party will be responsible for producing the first draft. In some common transactions there are conventions:

(a) In conveyancing, the seller's solicitor will usually prepare the draft contract, whilst the buyer's solicitor will prepare the draft conveyance or transfer.

(b) In share purchase transactions, normally the purchaser's solicitor will prepare the agreement and deed of indemnity and, by incorporating various warranties, require disclosure by the vending shareholders.

(c) In a leasehold context, it will be the landlord's solicitor who will prepare the lease and any licences.

In other matters the parties should agree where the responsibility for the production of the first draft lies.

4.6 Getting down to drafting

Once your preparation is complete, you can begin to draft. Try to find a quiet place and period in which you can devote time to producing a coherent complete draft. Trying to do so on the run in five-minute bursts between fielding phone calls, seeing clients and appearing in court will only produce certain disaster, it being most unlikely that a cohesive draft will result.

Shut your office door or find a vacant room. Ask your secretary, telephonist and receptionist to ensure you are not disturbed.

Before drafting substantively, you should prepare a skeleton of the agreement to ensure that you do not omit any material facts or legal points.

Where there are complex factual or legal questions, a skeleton will enable you to see the coherent and logical whole at the outset.

There are various opinions on whether the drafting should be done by hand on a fresh piece of paper, directly on to a PC, by dictation or by marking up a precedent. Melville in *The Draftsman's Handbook*, recommends the first course of action, including copying out any precedent clauses used, as this will force you to consider every word as you proceed, enable you to see the draft emerging, and will also enable you to break off at any point to research a further point. Shurman in *The Practical Skills of the Solicitor*, recommends dictation, as this will have the merit of speed, although he admits that complex clauses should first be written out. Melville rejects this approach as not permitting any of the benefits of handwriting the draft.

It is probably most common to mark up a precedent, especially if the document is already on the know-how system, or to draft directly using your PC. Using the basic photocopy or draft:

(a) Amendments should be clearly inserted by coloured pen.

(b) Deletions should also be clearly indicated by a single line being drawn through the text to be deleted.

(c) Any new clauses should be written out in whole or marked up from a photocopy of a precedent and firmly attached to the main draft either as a marked and numbered rider or by being stapled to the relevant page and clearly indicated where in the draft the clause is to be inserted.

If a word or few words only are to be inserted, do so above the line with an arrow indicating the position of insertion. If a long phrase or sentence is to be inserted, it is generally clearer to set the phrase out in the margin and mark it with an asterisk or cross or circle and indicate the place of insertion by marking the text with an arrow and the appropriate asterisk or mark.

It is worth doing this, even if you will be creating the draft on your PC.

Remember that if you are not creating the draft on a PC yourself, your draft will be transcribed on to a PC by someone to whom the sense of the draft may not be clear. Make sure that your draft is clearly legible and that crucial words such as 'not' cannot be confused with 'now'. If your writing is difficult to read, try drafting double-spaced, spell clearly in capitals any names or unusual words, and use blue, red or black pen and not some other exotic colour or pencil.

If using Microsoft Word, you should get into the habit of using the 'Track Changes' feature. Track Changes is a way for Microsoft Word to keep track of the changes you make to a document. You can then choose to accept or reject those changes. To turn on track changes hold down Ctrl + Shift + E or use the control in the Tools header. If you view the draft in 'Final Showing Mark-up' format, you can see all additions and deletions. Other document formats can allow similar comparisons to be made.

Be careful. If you are going to e-mail your document, you may not want other people to see your reviews and amendments, which might be readable by using the track changes features.

4.7 Appearance, style and content of the draft

Print your draft double-spaced on A4 paper.

Leave a generous margin each side, and at the top and bottom. This profligacy in paper will be seen as worthwhile when the ease of making amendments is contrasted with the prospect of amending a document submitted in single spacing. Such is the perversity of human nature that faced with the latter document one is more inclined to carp at the draft.

4.7.1 Does the draft form a consistent and coherent whole?

Generally, you should give each clause an appropriate heading, especially in drafting a lease or a commercial agreement, and number each clause and sub-clause consecutively. For no good reason conveyancers prefer to number clauses and define sub-clauses in brackets, e.g., clause 2(23), whereas company and commercial lawyers prefer to number clauses and sub-clauses decimally, e.g., 2.1, 2.1.1. Provided the numbering (or even lettering) is consistent and internal references correct, adopt the style you find most comfortable.

Numbering clauses also assists compilation of an index to the document. Beware, as this will often not be amended when the substantive document is being negotiated and the final document may be engrossed with an incorrect index.

The draft should follow your skeleton outline which you prepared from your client's instructions. Try to ensure that the clauses drafted from scratch fit stylistically with the precedent used as a base. You should ensure that the words and phrases defined in the draft are consistent throughout, for example, 'seller' and 'buyer' should not be mixed with 'vendor' and 'purchaser'. Try to use clear plain English.

Once drafted, read through the document carefully to make sure that clauses do not conflict with each other. This is especially important if you are not familiar with the precedent you are using as a base.

4.7.2 Is the draft logically organised?

4.7.2.1 Prescribed contents

To a large extent, the contents of the draft document will be dictated by the subject matter. Sometimes the contents will be prescribed by legislation:

(a) a transfer should comply with the Land Registration Rules 2003;

(b) a statement of case should comply with the Civil Procedure Rules 1998 (as amended);

(c) a statutory demand should comply with the Insolvency Rules 1986.

On the other hand a partnership agreement, commercial agreement or trust deed will follow no prescribed format save for the basis provided by any precedent used.

4.7.2.2 Opening clauses

Traditionally, a document starts with its own description, for example, 'This Deed' or 'This Agreement', followed by the date to be inserted, although the date of a Will is traditionally at the end. The parties to the document are generally then described and defined for ready reference throughout the draft. It is often better to describe them by their legal function in the document, e.g., 'Vendor', 'Purchaser', rather than by a shortening of
their names, e.g., 'Jones', 'Brown', which is often a little functional especially when the client reads the document through.

Try to identify the parties clearly as any inaccuracy may have repercussions, for example, leading to a search in the Central Land Charges Registry against an incorrect name in a conveyancing transaction. You should ensure that you refer to each party's full name and address, so that notices can be served correctly in future. In addition, you should ensure that every person who is entitled under an agreement is a party to that agreement, so as to avoid problems enforcing it through the courts. A company should be identified by its registered number as well as name to avoid any doubt if there is a subsequent change of name.

4.7.2.3 Definitions

In a commercial agreement the next clause is usually a list of definitions of terms used in the document. This is particularly prevalent in commercial leases and share purchase agreements, in which it is particularly important to define concepts such as 'Demised Premises' or 'Service Charge', the full impact of which can only be discovered when reading the substantive document. It is important to be aware of statutory definitions such as those contained in the Interpretation Act 1978, the Law of Property Act 1925, the Land Registration Act 2002, the Administration of Estates Act 1925 and the Companies Act 2006. These definitions can be easily incorporated into your draft.

4.7.2.4 Recitals

Other recitals may then follow giving details of the history or purpose of the transaction or reciting an agreement pursuant to which the parties are entering into a deed. In conveyancing, recitals assumed an importance (now reduced) under s. 45(6) of the Law of Property Act 1925. In a share purchase agreement details of the target company are often contained in recitals.

4.7.2.5 Operative clauses

Following any recitals, the main agreement or operative part of the document should be set out and this will be dictated by subject matter.

Do not forget that you are preparing a document for your client. Its purpose is to protect the client, but it will also generally have to be agreed by the other party. So be careful to tread a fine line, protecting your client whilst not placing intolerable conditions on the other party.

After all, agreement is usually the aim and objective of all parties. It is not your job, unless so instructed, to throw insuperable obstacles into the path to accord.

4.7.2.6 Impact of statutory provisions

Consider any matters which your analysis of the law may have pointed up, for example:

(a) In a trust deed, ss. 31 and 32 of the Trustee Act 1925, relating to powers of maintenance and advancement, will apply unless there is an express contrary provision.

(b) In a contract for services consider the terms implied by the Supply of Goods and Services Act 1982, ss. 13 to 15, and any prospect of contracting out.

(c) In a consumer credit agreement, be aware of the extortionate credit bargain provisions under the Consumer Credit Act 1974, ss. 137 to 139.

(d) It is not possible to contract out of liability for death or personal injury (Unfair Contract Terms Act 1977, s. 2(1)).

(e) In a lease, if consent to an assignment may be given by the landlord, then, under the Landlord and Tenant Act 1927, s. 19, such consent may not be unreasonably withheld.

4.7.2.7 Contingencies

After setting out the parties' relevant obligations and rights, there may be matters of a 'what if' nature which should be incorporated. Typically in a share purchase agreement there are many provisions dealing with eventualities if the vendors are in breach of warranty. These would deal with matters limiting liability, conduct of claims and any taxation adjustments. It is not always possible to foresee and make provision for every consequence arising from the parties' subsequent dealings, but, when taking instructions and advising on the law, certain matters should be raised with the client, for example:

(a) in a will, to make provision in the event of beneficiaries predeceasing or there being a partial or total intestacy;

(b) in a lease, to make provision for the consequences of the tenant paying rent late;

(c) in a partnership deed, to make provision for the status of the partnership and its assets in the event of a partner retiring or dying;

(d) in an option agreement, to make provision in the event of the option not being exercised;

(e) in a joint venture agreement, to make provision if either party is in breach of its obligations;

(f) in a mortgage, to make provision in the event of the mortgagor defaulting.

The above list is by no means exhaustive and the matters to be considered will vary depending on the client's instructions and the solicitor's own experience. Many precedents will deal with certain events as a basic checklist, whereas others may have special clauses to deal with them.

4.7.2.8 Standard clauses

Many commercial agreements contain standard, or 'boilerplate', clauses. These will cover such matters as:

(a) no partnership between the parties;

(b) jurisdiction;

(c) entire agreement;

(d) certificate of value;

(e) arbitration in the event of disputes;

(f) service of notices.

4.7.2.9 Signature

The main body of the document is brought to an end by reciting the signatures to be appended. If the signatory is under a disability or an attorney is signing, the appropriate testimonium and attestation clauses should be affixed. A person who is named as more than one party should execute the document once only.

For provisions relating to execution in special circumstances, see the *Encyclopaedia of Forms and Precedents*, 5th ed. vol. 12, pp. 369–72. Check also any special requirements for witnessing. Certain statutes and statutory instruments require signatures to be witnessed.

4.7.2.10 Schedules

The modern tendency in drafting, one brought on by word processing, is for the operative part of the agreement to be concise, but to refer to lengthy schedules containing the bulk of the agreement. In considering whether to schedule or not to schedule, you should consider whether removing a discrete part of the agreement to a schedule will aid or hinder the logical organisation of the whole, and whether it will render the agreement any more or less coherent.

The way in which a commercial lease is structured nowadays is a good example of the use of schedules to clarify a very long document. The rent review clause became the first to find its way into a separate schedule. This was logical as it concerned a special procedure occurring outside the main leasehold relationship. Shortly after, in leases of shopping precinct units, the description of the services being offered or made available or charged for in the service charge formed a second schedule. This was an attempt to keep the reader's interest in the important mechanics of the service charge and not to deflect it with a two-page recitation of services in the main body of the lease. Also, it is less likely that a list of services in a schedule would be amended than if it were in the main body of the lease.

As these shopping precinct leases were prepared in bulk for large developments, the next step was to remove the description of the demised unit and any rights and reservations from the main body into another schedule, which was the only part that needed to be altered on the word processor of the lessor's solicitors. From there the floodgates opened, and more or less discrete areas are finding their own schedules, so that some leases now have an operative part which does no more than incorporate the schedules.

It must be stressed that it is not wrong for that to be the final outcome, so long as you are happy with the draft, the client is happy with the draft, and the draft meets the client's goals. What must be stressed is that any schedule must be properly incorporated into the agreement.

4.7.2.11 Checklist of clauses

(a) Date;

(b) Specification that this is a deed;

(c) Names and addresses of everyone signing;

(d) Necessary definitions;

(e) Consideration, payment and receipt;

(f) Main agreement;

(g) Warranties, conditions, covenants;

(h) Boilerplate clauses;

(i) Signature clauses;

(j) Schedules;

(k) Plans.

4.8 Your draft in their hands

4.8.1 Consideration of your draft

Many people may pore over the fruits of your labours. Chief amongst these must be you, yourself and nobody else checking that the draft meets the client's objectives. Moving on, the draft may also be looked at by your secretary, your principal or partner, the client, the solicitor on the other side, the other party, and possibly the solicitor on the other side's secretary. You may hope that this is as far as it goes—though there must be at the back of your mind the potential for disagreement and litigation, when your drafts may be considered by other solicitors, counsel and the judge.

4.8.2 Your secretary

In most firms it will be likely that you will prepare the draft yourself on your PC. However, if you do not, you will be asking someone else to prepare it for you. Try to explain the draft document you are asking them to prepare. A few minutes at the outset explaining names and the purpose of the document may save hours of later frustration. It may enable the secretary to make educated guesses in the event of doubt or your illegible handwriting rather than to chase you up.

Try to explain genuinely when you require it. Everything in practice is always 'urgent', and saying so makes no difference. If you have a deadline, make it known at the outset. If the deadline is one of enormous urgency, do not expect perfection to be handed back to you.

Policy differs from firm to firm. Everyone should take pride in their work, but some firms may prefer work to be turned over swiftly at the expense of accuracy. Whatever the policy, it is up to you to check carefully and proofread the draft produced on the word processor not only for spelling but also against your handwritten or dictated draft. Also it is worth reading the draft as if it were a document sent to you by someone else, to see whether it still makes sense to you.

If there are errors in what has been handed to you, clearly indicate on the typed draft the amendments to be made, preferably in red pen, and indicate insertions, deletions and spelling corrections. You should ensure that the typed draft is dated and the amended typed draft is also dated. The date may be automatically generated by the word processor, but it does not hurt to indicate somewhere the draft number and the date it is created.

Once it is typed correctly, keep a copy of the draft on the file and do not write on it. You should as a matter of course keep all drafts in order on the file even when they are superseded. Remember to note clearly which version a draft is.

The document management system in your firm ought to keep a record of all drafts. But it is important that you can distinguish in future what your original draft was.

Summary

(a) Remember: you are legally trained; your secretary may not be.

(b) Write or dictate clearly.

(c) Explain the draft.

(d) Check what is typed carefully. It is best if you can allow yourself time and distance from your preparation of the draft and checking it. If you are too close to the original, your brain will read what you thought you drafted rather than what is on the paper.

(e) Try to read the draft over with a colleague.

4.8.3 Your principal or partner

Generally as a trainee solicitor, you will be closely supervised and your principal or supervising partner should check your drafting closely. As with any skill, you can always pick up tips from those around you and ideally, time permitting, your draft should be considered together by you and the supervisor. If that is not possible, ask for the draft to be returned to you with amendments and comments so that you can digest them.

Note those points which are substantive, and ask yourself why they were omitted or incorrect in your draft.

Note the points of style and consider whether these are an improvement on your own.

Do not automatically adopt your principal's style. One style is not better than another; everyone should use the style with which they feel most comfortable. Your principal's style will have been developed over many years.

Always try to make time to discuss matters about your drafting with your principal and ask for regular feedback. As with everything, familiarity with the subject matter of a type of document will lead to an improvement in your drafting.

4.8.4 Your client

As has been emphasised before, a solicitor produces documentation at the request and expense of the client. It is therefore sensible to send a draft of an agreement to the client for comment. It is not for the client to check that you have got the agreement right; it is expected that you will have the law correct. This is the opportunity for the client to give you any missing pieces of information that may not have surfaced at the interview, or for you to ask the client what is to occur in a eventuality to which you have not previously drawn the client's attention.

How this process occurs often depends on the client and the nature of the documentation. If the client is sophisticated, used to reading documents drawn by your firm and familiar with the type of transaction, it is more sensible to send the draft with a few pointed comments. On the other hand, if the document and/or the legal language is likely to be unclear, it may be better to discuss the document in a meeting with the client.

Of course, not every document needs to be checked with the client before submission to the other side. It would be unnecessary, for example, for the client to see a standard conveyance. On the other hand, the client should always have an opportunity to see and comment on a draft Will.

4.8.5 The solicitor on the other side

Drafting in the non-contentious sphere is mainly a collaborative exercise, enabling parties to come to an agreement. This suggests a meeting of minds, a fair outcome, and relative equality of bargaining power. Although this is the case, the process is often seen by the participants in confrontational terms, with each side scoring points off the other. Presumably, this is because most people perceive the legal process to be a confrontational system with the public image being of the courtroom rather than of the drafter or 'conflict blocker'. Hence, the other party is described often as being 'on the other side'.

It must be remembered that your client will want an identical outcome to the other solicitor's client, provided that there is no detriment to either.

Take a domestic conveyancing situation. Your client wants to buy a house and move in. The other party wants to sell and move out. Both parties should have an identity of interest. There will be almost inevitable confrontation between the solicitors over the rate of interest for late completion. It is too high for one solicitor and too low for the other. Never mind that neither client has the slightest intention to complete other than on the fixed date. This confrontational aspect can then all too often spill over into the whole transaction.

So it is best to try to avoid unnecessary discord which may be a bar to agreement. Usually, within reason, the final agreement entered into will be fair to both parties, and there is little to be gained by preparing a first draft that is unduly restrictive and one-sided. Time, money and effort can be saved if the first draft prepared can be used as a basis for final agreement.

Remember always, though, you are acting for your client and on your client's instructions and your primary duty is to protect your client's position.

Mention was made above about equality of bargaining power. You may find yourself acting in cases where there is clearly an inequality of bargaining power. In such cases, especially when preparing contracts in the entertainment industry, you must warn your client of the possible consequences of any agreement which is one-sided or restrictive being overturned.

So, once you have instructions from your client to submit the documentation to the other solicitor, it is courteous to do so by sending clean word-processed copies in duplicate, thus providing a file copy for the other solicitor. The drafts should be accompanied by a letter and, if appropriate, any comment on unusual provisions should be made there. If the reasons for such a provision can be seen from the outset, it is less likely that wholesale revision will be proposed.

One of the drafts submitted can then be used as the travelling draft by the other solicitor and returned to you with any amendments. It may be disconcerting to find one morning that your pristine draft agreement, over which you slaved for many hours, has been returned with red amendments on every page. Suppress the natural reaction of outrage and remember that the other solicitor is not being confrontational (even if he is) but merely putting his client's input to reaching a final agreement.

Consider the amendments carefully, and see if they are not a help to your own drafting skills.

How are points of substance worded? Are there any stylistic amendments?

Has the other solicitor spotted any inconsistencies or imprecision in your draft which you failed to pick up?

4.8.6 The other solicitor's client

Remember that just as you would refer your draft to your client for confirmation of instructions, expect the solicitor receiving your draft to send a copy to his or her client for comment. As a matter of common sense, do not refer in derogatory terms to the other party or word any provisions in inflammatory terms which would detract from finding a final agreement.

For example, in a mortgage it would be preferable not to define one party as 'the Debtor' and the other as 'the Creditor'. It is better to use 'the Mortgagor' and 'the Mortgagee'.

4.8.7 The solicitor on the other side's secretary

In commercial matters, the process of amendment often occurs not by use of the travelling draft, but by retyping. If this happens, carefully check the retyped amended version sent to you, not only to see the amendments but also to detect any transposition or omission of your original text. Remember, any error your office can make can also be made anywhere else. Agree whether a tracked changes facility will be used. This should enable you to see by means of different colours the author of any reviews.

4.8.8 Others in the event of dispute

This is the situation which by your skill you are trying to avoid: the time when the parties are in dispute over the provisions of the document you prepared. This may also lead to accusations that you did not produce what your client wanted.

A good agreement should be watertight to prevent litigation. Even if it is, this is no bar to litigation. Often, the litigation may take place many years after the agreement was negotiated.

To be able to deal with litigation when it occurs, make sure at the time of drafting that your file is in good order. Keep:

(a) your first draft and any precedents you used as a basis,

(b) the first word-processed draft,

(c) a draft incorporating your client's comments,

(d) the draft as submitted to the other side,

(e) the travelling draft with all comments or amendments,

(f) all word-processed versions as redrafted,

(g) the final version as approved by your client, and

(h) a photocopy of the completed document.

Whilst this may appear to be unnecessarily bulky, it is far more useful than relying on your recollection. These drafts can prove that you proposed amendments, which were rejected, and that your client approved the final draft without your amendment. It can also assist in rectifying any mistakes.

4.9 Use of grammar and language

A lawyer must be skilled at communication to be truly effective. You must be able to interpret the facts of a matter to fit into a legal context and communicate the legal conclusions to the client. The interview, negotiation and advocacy all depend on verbal communication. Drafting and legal writing depend on written communication, which relies upon the written word solely to convey meaning.

Verbal communication encompasses more than reliance on words. Tone, volume, inflection and body language, for example, add greatly to the listener's understanding of the words. On the page, the words stand naked, shorn of all nuance, save for their direct meaning and context. Written communication needs to be more precise than its verbal counterpart.

Take the words 'John did it'. In speech, this can be an accusation, an expression of surprise, an expression of delight, a questioning phrase, or a statement of fact, depending on inflection and emphasis. On the page, one cannot be sure of the precise meaning to be conveyed without more of the context.

'John did it', he yelled as the winning runs sped to the boundary.

Now at least the words are in some context, but it is still not clear whether they show surprise or delight.

Words written on paper require interpretation and if the interpretation is open to ambiguity there may be potential for dispute. Part of the lawyer's drafting skill is to ensure the document prepared means what the client has instructed the lawyer to prepare.

The concerns for writing plain English, using proper grammatical constructions and choosing appropriate language detailed earlier in the context of legal writing (**Chapter 3**) are equally applicable to the formal agreement.

EXERCISE 4.5

Are there any barriers to using plain English in a legal draft? List any advantages and disadvantages both legal and practical in using plain English.

You must always bear in mind the subject matter of the document being prepared. There is a certain amount of formal necessity in many agreements and deeds, where wording, even though it may seem arcane, has been judicially or statutorily approved.

That is one of the purposes and comforts in using precedents. If the words have already been approved by the courts, then if used again in the same context they should bear the same meaning.

It is not wise to depart from the approved wording simply because you prefer a plainer style, as your paraphrase or rewording may be ambiguous and uncertain where before there was legal clarity and certainty. Take, for example, the forfeiture provision in the Rugby Borough Council tenancy agreement commented on in [1984] Conv 325. Here the plain English does not convey the legal meaning of the original and it also falls foul of the problems raised by s. 210 of the Common Law Procedure Act 1852.

Traditional drafting eschewed punctuation for no good reason. It appears that this was a consequence of Acts of Parliament being passed in manuscript without any punctuation, and thus judicial interpretation had to disregard any subsequent punctuation inserted by printers. See, for example, *Commissioners of Inland Revenue* v *Hinchy* [1960] AC 748. Modern drafting should be properly punctuated with appropriate use of comma, full stop, colon and semicolon. For a discussion of proper usage of punctuation see Melville, *The Draftsman's Handbook* or *The New Fowler's Modern English Usage*.

Beware incorrect punctuation, which is worse than no punctuation at all. Inadvertently, a comma may parenthesise a clause that was meant to be read conjunctively, rendering a wholly unexpected meaning.

4.10 The process of amendment

4.10.1 Deciding what amendments to make

We have already considered the question of production of a draft document and submitting it for consideration. Unless the document is solely for the client, such as a Will, it is likely that it will be considered on behalf of another party. Almost invariably there will be some amendment.

The art of amending documentation is part of the skill of drafting and can be seen as being analogous to the skill of adapting a precedent. In this case, however, the precedent is the draft document into which you have to fit your client's instructions.

On receiving a draft document you will need full instructions from your client to provide the context in which the draft must be read. This may of course be obvious and you may be able to read the draft and consider amendments before you have full instructions, but you cannot carry out the task of amendment without full instructions any more than you can the task of drafting.

Secondly, read the draft document supplied. Do not assume that as you are familiar with documents of this type, the draft will contain provisions in the terms in which you expect to see them. Do not assume, if you have already received similar documents from the same firm or same client, that the draft will be the same as that previously supplied. In courtesy, the firm submitting the draft ought to point out amendments or similarities to the draft previously supplied, but do not assume anything from the absence of comment. Your duty is to your client and you must read the draft carefully. Read not only what is included in the draft, but consider what has been omitted. If the source of the draft is clear, return to the original precedent and look at what has been amended or omitted.

Thirdly, go through the draft with your client pointing out and explaining the provisions and omissions. Take instructions on every point and note those instructions carefully. Advise the client on any points or clauses you would propose to amend and the reasons for amending. Does the client agree? Take instructions on the strength of feeling that the client has for the amendments in case they are not accepted. Suggest possible compromises which you might propose in the event of your amendment being rejected.

Often this will reveal three types of amendment to be made:

(a) matters of principle relating to the transaction which your client understood would be incorporated, but have been omitted or not incorporated in the terms the client agreed;

(b) matters to which you as a lawyer cannot permit your client to agree, as these would not afford appropriate protection in the transaction;

(c) matters of flexibility to which both sides will readily compromise, but which are used as negotiation tools or initial 'try-ons', being matters not previously discussed or agreed, but ones on which you endeavour to ensure favourable treatment.

Each of these categories of amendment should be dealt with by you with equal precision, but in negotiating the finally agreed document they will receive different emphasis.

Take, for example, a commercial lease. Consideration of the document with your client reveals that interest on a late payment of rent is at 5 per cent above base rate, the demised premises do not have a specific right of access over stairs to them, and a provision entitling the client to pay only 50 per cent of the service charge has been omitted.

Your client should in instructions inform you of the agreement on the service charge, which will no doubt have been one of the factors influencing the decision to take a lease of the premises. It is thus very important to make and insist upon the appropriate amendment. If indeed this is a matter that the parties have agreed upon, your amendment will be readily accepted. If it is not, clearly it is a matter for further negotiations by your client.

No doubt your client would be grateful if you could reduce the interest rate in the event of rent default. But if you are unable to achieve a reduction, this is unlikely to affect the decision to proceed. No doubt the landlord will be surprised that you would seek to amend such a provision. You would probably suggest 3 per cent over base rate and compromise at 4 per cent over base rate. Such a compromise can be seen as a hard-fought point sacrificed to achieve a more meaningful agreement elsewhere.

The absence of any legal right of way to the premises would also probably not concern the client unduly if it were clear on the ground that access could be had without obstruction. But this is where the client relies on your expertise, and you must stress the importance of such a defect and of insisting upon a proper solution to the problem in the completed document.

4.10.2 Making amendments

Essentially there are five ways of making amendments and negotiating documentation. These are:

(a) by travelling draft (whether in hard copy or by electronic copy);

(b) by retyping (now usually not used);

(c) by letter/fax/e-mail (with or without attachment—such attachment typically being the travelling draft);

(d) by telephone or meeting negotiation;

(e) by disk (now again normally replaced by the e-mailed travelling draft).

The traditional method favoured by conveyancers is for one copy of the draft document, the travelling draft, to be sent from firm to firm with subsequent amendments shown by coloured pen. The first amendments are generally shown in red pen, the second in green, the third in violet and the fourth in yellow. This is the same convention as when amending statements of case in litigation. Other amendments are made in whichever colour is most readily to hand. Whilst this process used to be done by a single hard copy being used and being sent from solicitor to solicitor by post, it is almost certainly now done by an e-mailed copy being sent with changes being evident by the colours shown up on the track changes feature.

Ensure, if using a hard copy, that an amendment by way of travelling draft is clearly written, showing precisely where the amendment is to be made, and not obscuring the original text. If possible, delete by a thin line and make an amendment in the margin. Also make any notes on amendments in the margin.

To reinstate a deletion it is traditional to mark 'stet' next to the deleted wording and put a broken underlining beneath the words to be retained.

Before returning the travelling draft, whether by hard copy or by email, it is wise to mark up or print off (with the changes easily visible) a copy for your file, which mirrors the travelling draft. When receiving the travelling draft, mark up your file copy with the other side's comment and amendment before reamending. It is important to remember that at the end of the day the travelling draft will be kept by the solicitor who is engrossing the final document and so the file copy must be kept by the other solicitor not only to check the final engrossment copy but also as a record in the event of later dispute.

Amendments, particularly if they are minor or small in number, may be made by letter, fax or email. The letter should refer clearly to the page number, clause and sub-clause number, and line number and quote the wording to be amended and indicate the amendment to be made. This can often be helpful as you can show your reasoning in context with the amendment. However, it may be awkward to read the amendments without the draft and there is room for subsequent error when transcribing the finally agreed amendments from correspondence to the draft. This method is often used in conveyancing, although the amendments may often be proposed in the form of preliminary enquiries and requisitions.

In some circumstances, it may be appropriate to propose and discuss amendments by way of negotiation over the telephone or in a meeting rather than by the transmission of written amendments. The process of substantive amendment is the same as in drafting, but the techniques involved are those of negotiation.

4.10.3 General principles

Whichever method of amendment and negotiation you adopt, make sure that both the content and the position of your amendments are obvious. Always check that your amendment has been considered.

When making amendments it is courteous to inform the receiving solicitor of the reason for amendment, especially if it is a matter of principle which your client believed was previously agreed.

Do not amend simply for stylistic reasons. Do not change designations simply because you prefer to use your own appellations.

If you delete a clause, ensure that any consequential renumbering is also made and that cross-references are suitably amended.

If there is more than one document under consideration, for example, in a share purchase with deeds of indemnity, schedules of warranties and option agreements, ensure that your amendment operates consistently throughout the documentation.

Confirm your amendments with your client and continue to seek appropriate instructions.

4.11 Engrossment and completion

Once a document has been agreed in draft by both parties, it will be ready to be entered into so as to create the appropriate legal relations. Generally speaking, the creator of the first draft will print out the final document for signature, which is known as the 'engrossment'. Originally engrossments were written or printed on vellum or parchment. Nowadays, the final print will be on stiff quality A4 paper.

Before the engrossment is run off, you should get a final draft which you should check word for word against the travelling draft and also consider for spelling and general coherence. Only if this conforms to the agreed version should you get an engrossment run off.

Invariably this document will consist of single A4 pages which need to be bound together. This can be achieved by the document being sewn with green ribbon, either in the corner or down one side, or by binding in a machine. These machines operate a method of heat binding, the efficacy of which can wear off over the years, especially if the document is handled frequently. Stapling is too easy to undo and pages are prone to becoming detached over the years. Sadly the art of sewing documents is dying out.

Check the engrossment then to ensure that all the pages are included in the right order, and check especially any exhibits or appendices or enclosures. Plans in leases also need to be incorporated into the engrossment. Make sure these are appropriately coloured before binding or sewing. Then ensure that the plan, if larger than A4 size, is folded properly so it can be opened and examined without being obscured by the binding.

It may be necessary to prepare duplicate engrossments. In any case you should also get a file draft copy of the unsigned engrossment to keep on your file. This should be dated, as with all the other drafts, and should mirror the engrossment in every way by inclusion of all appendices, plans or enclosures.

Signature may take place in a meeting of all parties or separately by one party before transmission of the documents to the other. If there is no meeting of parties, the client may sign in the presence of the solicitor or the solicitor may send the document by post to the client.

If the signing takes place at a completion meeting, you should go through the final document with your client to make sure that it corresponds with instructions, that the client understands the rights and obligations created by the document, and that the client is satisfied. If the document is not prepared for the completion meeting by yourself, you should read the final version carefully through to check that it conforms with your understanding of the negotiated draft document, from your marked-up file copy, and that no subsequent amendments or clauses have been 'sneaked' in by the other side. Many a (no doubt apocryphal) story is told of the completion meeting taking place long into the night where the lawyers were drafting in one room and a secretary typing up the draft in another, and so as not to break matters up the client put himself up as go-between, and once out of the room persuaded the secretary to make amendments of his own volition unbeknown to the lawyers or the other party. This subterfuge, so the story runs, is then only discovered by the diligent lawyer, having worked through the night, taking the trouble to recheck the whole document before allowing signature.

If the documents are signed separately, make sure they are not delivered or dated until they are finally agreed and properly executed. It is always preferable that signature occurs with the lawyer present, but this may not always be possible, especially when companies execute documents. You should then explain carefully the formal requirements for signing and, if necessary, the requirements for provision of witnesses, and any instructions for dating the document. Following the correct procedure for execution of a Will is of crucial importance.

Any last-minute alterations should be clearly marked on the engrossment by hand and all the parties (including any witnesses) should initial the alterations in the margin. Some lawyers ensure that the parties initial every page.

The document will come into effect once it has been completely executed and dated. You should keep a photocopy on your file for record or make up your engrossment file copy by filling in the date and details of the signatures. A copy should also be sent to the client, especially if the original is to be held elsewhere than by the client.

Do not omit to deal with any post-completion matters promptly. These include payment of stamp duty, registration with the appropriate registry, or filing at court.

Summary

(a) Carefully proofread the final draft before engrossing.

(b) Ensure that the document is appropriately engrossed and bound.

(c) Be clear about who signs the document when and where.

(d) Keep a copy of the completed document.

(e) Just because the document is signed does not mean that there is nothing more for the lawyer to do.

4.12 Construction of documents

The aim is to prepare documentation which covers all the requirements of the legal relationship between the parties, and so clearly that all dispute can be resolved by reference to the final agreement. Unfortunately, this does not always work out. So it is worth bearing in mind when drafting some of the important rules of construction which the courts use to interpret agreements.

The primary consideration in construction is what was the intention of the parties. Donaldson J said in *Segovia Compagnia Naviera SA* v *R. Pagnan & Fratelli* [1975] 2 Lloyd's Rep 216: 'The duty of the court is to ascertain the presumed common intention of the parties, to be deduced from the words used and the background to the transaction. Their actual, but uncommunicated, intentions are irrelevant.'

If there is a conflict between the printed or typed word and the written word the latter is to be preferred to the former.

Any ambiguous term is to be construed most strongly against the party for whose benefit it is intended. This is known as the *contra proferentem* rule. See, for example, *Burton and Co.* v *English and Co.* (1883) 12 QBD 218.

Under the *eiusdem generis* rule, specific words limit the meaning of general words following to the class of the specific. But if the general words lead to the specific, no such limitation applies. See, for example, *SS Knutsford Ltd* v *Tillmans & Co.* [1908] AC 406.

The document must be looked at as a whole. This is an extension of the rule concerning the parties' intentions. See, for example, *Nereide SpA di Navigazione* v *Bulk Oil International Ltd* [1981] 3 All ER 737.

The document should be given its grammatical construction unless there is an expressed contrary intention.

Words will be given their ordinary dictionary meaning, but technical words will be given their technical meaning. However, the meaning must be limited by the context in which the words are placed. If there is an obvious error, the court can correct it in the way a reasonable man would expect.

Where words are capable of bearing two constructions, the reasonable construction is preferable.

Words repeated in the same document will be given the same meaning throughout.

An express term will override an inconsistent implied term.

4.13 Plan, write, revise

The theory of 'plan, write, revise' applies in the context of drafting legal documents as it does in writing legal letters. This three-part analysis of the way efficient writers, especially lawyers, work is a helpful aid in breaking down the process into its constituent elements.

In the planning stage, the lawyer must:

(a) take instructions;

(b) analyse those instructions and compartmentalise them in the factual context and in the legal context;

(c) ascertain the purpose and objective behind creating the document;

(d) research the relevant up-to-date law and find any appropriate precedent.

In the writing stage, the lawyer must:

(a) prepare any skeleton draft;

(b) create ideas for the draft;

(c) concentrate on ensuring that all the content required after the planning stage is incorporated into the draft document.

In the revision stage, the lawyer must:

(a) re-analyse the instructions, factual situation and legal research;

(b) rewrite to ensure coherence;

(c) reorganise the material in a clear and user-friendly manner;

(d) edit the material;

(e) concentrate on spelling and grammar.

4.14 Persuasive and informative drafting

4.14.1 Types of legal drafting

Some legal drafting is designed to be persuasive and will require particular attention to be paid to selecting and organising the content. The layout of a persuasive draft is often dictated by the line of argument intended to emanate from the document.

Some legal drafting is designed to be informative only. Many letters reporting to the client fall into this category. The draft should contain all the information necessary and useful to the client. Often the tone of such a document will be neutral and unemotional, without any line of argument dictating the format. In such cases the draft will be organised in a manner that is most appropriate to the user.

Drafting agreements is done to create specific legal rights between the parties to the document, which the parties have freely agreed on. Legal documents of such a nature, which are by and large non-contentious, are thus informative rather than persuasive.

An informative document needs to be a self-contained whole. Just as when reading a good science-fiction novel, you are carried along if the author has created a world in which all the activities of the characters are rational when referred to that world and in which all the answers can be found by looking into that world, so the lawyer drafting the non-contentious agreement must ensure that the document is fully self-referential and covers all appropriate circumstances emanating from the rights and duties created by the document.

4.14.2 Drafting in particular areas of law

By looking in brief at some of the drafting expectations in the compulsory and core subjects on the Legal Practice Course, it is possible to point up some common and some unique points.

4.14.2.1 Drafting in conveyancing

In this subject, the following documents are likely to be met:

(a) the contract for sale;
(b) the conveyance;
(c) the transfer;
(d) the lease;
(e) the licence to assign.

The purpose behind each of the above is informative. They all create or transfer legal rights and in some cases, particularly the lease and licence, create legal duties. Little in the way of original thought is required from the drafter, there being specific statutory requirements for the contents of these documents. Contracts often incorporate many standard terms and conditions, which in themselves create and define the world in which the contracting parties find themselves. Conveyances and transfers incorporate statutory provisions.

It is only when drafting a lease that there is greater scope for expanding or contracting this universe of rights, duties and consequences, but often creativity is eschewed and trusted formulae or precedents are preferred.

4.14.2.2 Wills, probate and administration

In this subject the documents most likely to be drafted are Wills. Just as in the conveyancing arena, the purpose behind the will is to make an effective disposition of the property of the testator. Thus the document is informative rather than persuasive, and should be drafted in the appropriate language in an appropriate order. The structure is determined by the underlying purpose of providing information, with the most important information—the core provisions disposing of the property—being given before the necessary administrative provisions, the what-if clauses which complete and flesh out the drafter's world.

4.14.2.3 Business law and practice

In this subject you may be asked to draft a multiplicity of documents, including:

(a) a partnership agreement;

(b) a shareholders' agreement;

(c) a debenture document;

(d) articles of association;

(e) share purchase documentation;

(f) distribution agreements.

All these documents are concerned with the creation of rights and duties of the parties and it is clear that their purpose is, as in the documents dealing with property rights and transfer, informative and not persuasive. These documents differ from property rights and transfer documents in that there is much greater emphasis on what-if scenarios and legislation and implied terms play less of a part than default property provisions implied at law.

The drafter creating an agreement in this commercial sphere must ensure that the whole of the world environment in which the rights and duties are to be exercised is present in the document.

4.14.2.4 Litigation

Documentation to be drafted in litigation includes statements of case, orders, affidavits and witness statements. Some may categorise instructions to counsel as being documentation to be drafted within the context of this part, but often such documents possess a greater similarity to the matters described in the section on legal writing (**Chapter 3**).

Unlike the non-contentious drafting purposes referred to above, the main purpose behind drafting in a litigation context is to persuade the court of the veracity or justice of your client's case. These documents do not seek to create an all-encompassing world in which rights and duties are created, in which administrative and default provisions are prospectively confronted by agreement.

The content and structure are often determined by the desire to order the events in such a way as to make one interpretation only possible—the interpretation in favour of one's client. The structure may often therefore be dictated by matters outside the document, a chronology, a viewpoint, a certain logic, rather than arising naturally from an internal logic within the draft itself.

That is not to say that the procedures for drafting contentious documents should differ from the basic plan, write and revise practice outlined in this chapter, just that the grammar, structure and content will have different purposes behind them.

4.15 Conclusion

Happily in *The Pirates of Penzance* Frederick's contract was brought to an end, whether by *force majeure* or frustration we are not told, and he can now take his place as husband to the lovely Mabel at the age of 21 and not at 84.

So you now should be able to set out to chart a course amongst the rocks and eddies which beset the legal drafter.

As stated by Philip Thomas in 'Legal skills and the use of ambiguity' (1991) NILQ 14: 'The aim of the legal document is the use of such words and grammar as are necessary to achieve a stated, preconceived goal: to capture the common intention of the parties in a manner enforceable through law.' This is the essence of your task as a skilled drafter and in striving for clarity or plain English, understanding or originality, style or form, you must not lose sight of the ultimate objective.

Just as with any lawyerly skill, you will profit from experience, gain confidence and find a style and rhythm of your own. However, it is hoped that you may profit from a consideration of some of the points in this chapter.

4.16 FURTHER READING

Atkin's Court Forms (London: Butterworths, 1999).

Blake, S., *A Practical Approach to Legal Advice and Drafting*, 5th ed. (London: Blackstone, 1997).

Encyclopaedia of Forms and Precedents (London: Butterworths, 1994).

Garner, B. A., *The Elements of Legal Style* (New York: Oxford University Press, 1991).

Making Good Solicitors (London: National Consumer Council, 2001).

Melville, L. W., *The Draftsman's Handbook*, 2nd ed. (London: Longman, 1991).

Plain English for Lawyers (London: National Consumer Council, 1984).

Shurman, L., *The Practical Skills of the Solicitor*, 2nd ed. (London: Longman Professional, 1985).

4.17 FURTHER EXERCISES

EXERCISE 4.6

Jones Ltd will supply expertise and crude refinery piping, while Harris plc will front the deal. Jones will receive an annual minimum payment of £2,750,000 to cover the expertise, but this payment will be increased if it is exceeded by the cost of the piping in any one year. Jones will supply a variable amount of piping each year (estimated to be between 1 million and 2 million metres per year) at an initial cost of £1.20 per metre, such price to be increased annually on the anniversary of the agreement by the RPI figure of inflation. You should think about what can be done if the RPI ceases to be published.

Draft the price clause.

Note: The RPI referred to in the problem is the Retail Prices Index, which is published monthly by the Office for National Statistics as an indicator of inflation.

The price equation should be P 5 OP 1 OP 3 (B 2 A)/B where P is the new price, OP is the original price, A is the RPI figure on day 1 and B is the RPI price on the review date. Thus if in year 1 the price is £6.00 and the RPI figure on the date of the agreement is 100 and the RPI figure on the review date is 120, the new price would be:

$$P = 6 + (6 \times (120 - 100)/120)$$
$$= 6 + (6 \times 1/6)$$
$$= 6 + 1$$

New price = £7.00

EXERCISE 4.7

Redraft the old provisions of the Trustee Act 1925, s. 36(1) and (7) so that the subsections are written in clear and plain English and all the provisions are accurately and concisely set out, with appropriate emphasis given to each provision. In particular, you should make appropriate use of punctuation.

36 *Power of appointing new or additional trustees.*

(1) *Where a trustee, either original or substituted, and whether appointed by a court or otherwise, is dead, or remains out of the United Kingdom for more than twelve months, or desires to be discharged from all or any of the trusts or powers reposed in or conferred on him, or refuses or is unfit to act therein, or is incapable of acting therein, or is an infant, then, subject to the restrictions imposed by this Act on the number of trustees,—*

(a) *the person or persons nominated for the purpose of appointing new trustees by the instrument, if any, creating the trust; or*

(b) *if there is no such person, or no such person able and willing to act, then the surviving or continuing trustees or trustee for the time being, or the personal representatives of the last surviving or continuing trustee;*

may, by writing, appoint one or more other persons (whether or not being the persons exercising the power) to be a trustee or trustees in the place of the trustee so deceased remaining out of the United Kingdom, desiring to be discharged, refusing, or being unfit or being incapable, or being an infant, as aforesaid.

(7) *Every new trustee appointed under this section as well before as after all the trust property becomes by law, or by assurance, or otherwise, vested in him, shall have the same powers, authorities, and discretions, and may in all respects act as if he had been originally appointed a trustee by the instrument, if any, creating the trust.*

5

Legal research

5.1 Learning outcomes

This chapter deals with the skills needed to carry out legal research. In this chapter we focus particularly on:

- understanding the importance of legal research;
- establishing an overall strategy in undertaking legal research;
- using paper-based and electronic resources to efficiently undertake research into a variety of focused and unfocused legal and factual problems.

5.2 Introduction

In the Legal Practice Course, the term 'legal research' carries a different meaning from that which is familiar to you from academic study. Practical legal research encompasses not just library-based research techniques, but the skills of legal analysis and fact investigation and management. These skills are particularly identified in **Chapter 6**, 'Practical problem-solving'.

5.3 The written standards

The Law Society has published written standards for the skill of Practical Legal Research. Under these you should understand the need for thorough investigation of the factual and legal issues involved in a client's matter, the need for preparation and the best way to undertake it. You should be able to:

(1) identify the objectives of the employer or client;
(2) identify and analyse the factual material;
(3) identify the legal context in which the factual issues arise;
(4) identify appropriate sources for investigating the relevant facts;
(5) determine when further facts are required;
(6) identify and analyse the legal issues;
(7) apply the relevant legal provisions to the facts;
(8) relate the central legal and factual issues to each other;

(9) identify the legal, factual and other issues presented by the documents;

(10) analyse the client's instructions and be able to identify the legal, factual and other issues presented by them; and

(11) record and present the results of research in a clear, useful and reliable form

You should also be able to demonstrate an understanding of:

(1) the use of primary and secondary sources;

(2) the methods of locating cases and statutes;

(3) the use of periodicals, digests and standard practitioner texts;

(4) the use of indices and citators; and

(5) the use of electronic research tools.

This section firstly contains an overview of the research process and methods of locating primary sources in diagrammatic and checklist format and is then followed by an expanded explanation of the process for doing library-based research, with a more in depth explanation of the types of resources. Finally, there are a number of exercises to use as practice.

5.4 The route to research

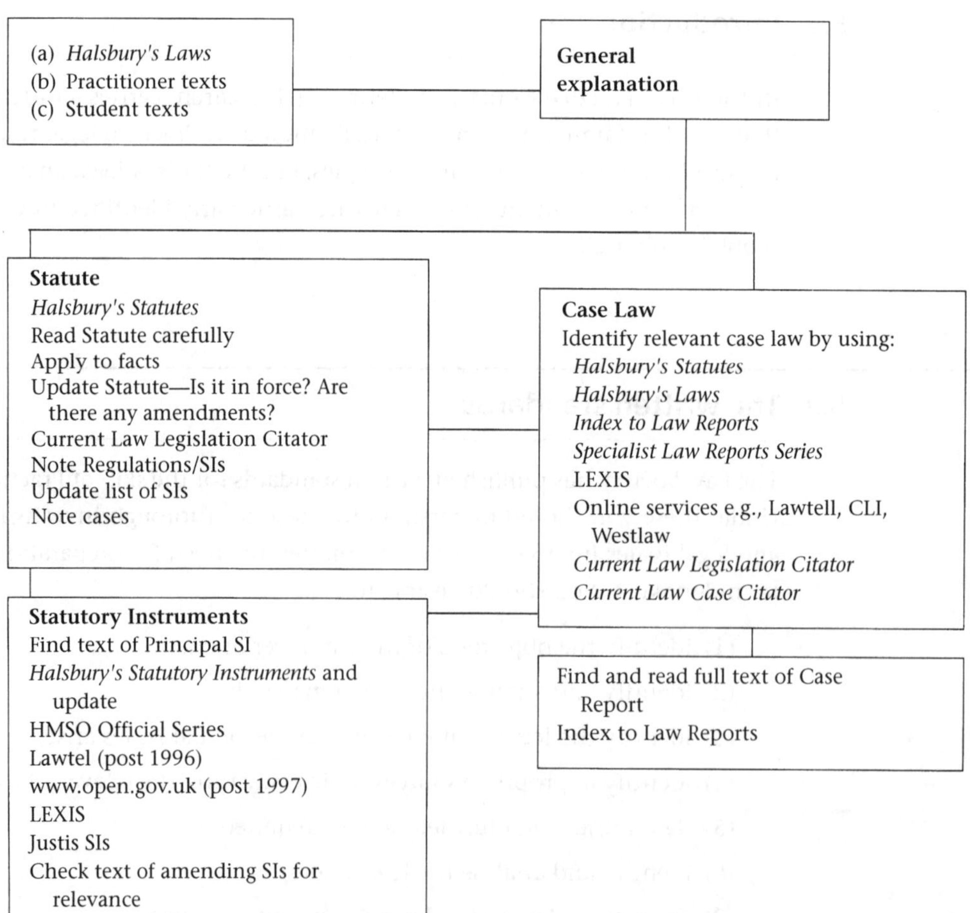

5.5 Using primary sources

5.5.1 Statutes

Finding a statute

Citations

(a) Inform Short Title: Land Registration Act 1925; or

(b) official citation: 1989 c. 12 (the 12th Act to receive Royal Assent in 1989) n.b. Acts receiving Royal Assent before 31 December 1962 use Regnal Year and Chapter number i.e., 4 & 5 Vict. c. 6 (the 6th Act to receive Royal Assent in the Parliamentary Session covering the 4th and 5th years of Queen Victoria's reign).

Search by using a citation

(a) *Halsbury's Statutes* (arranged by subjects);

(b) *Current Law Statutes Annotated* (arranged by year);

(c) Law Reports: Statutes (arranged by year);

(d) Public General Acts and Measures (by year);

(e) Internet at www.open.gov.uk (but only post 1988); or

(f) LEXIS or Westlaw (each section of an Act is separately listed).

Search by subject

Use the indexes, alphabetical lists, tables of contents:

(a) *Halsburys' Statutes* or *Halsbury's Laws;*

(b) *Index to the Statutes* (up to 1990);

(c) *Current Law Legislation Citator;*

(d) Textbooks and Practitioner Encyclopedias; or

(e) LEXIS.

Updating

(a) check supplements, continuation or replacement volumes, current services, loose parts and noters-up to statute series to determine if the statute is still in force, or if it has been amended, or repealed in full, or in part;

(b) look up the statute in relevant volumes and monthly issues of Current *Law Legislation Citator and the Current Law Monthly Digest;*

(c) check weekly journals e.g., *New Law Journal, Solicitor's Journal;*

(d) check on-line using Current Legal Information, LAWTEL, Westlaw and LEXIS; and

(e) very recent changes can be found by using House of Commons Weekly Bulletin as well as the weekly journals and on-line sources.

Finding an older statute

(a) Statutes at Large (From Magna Carta to 1859);

(b) Law Reports Statutes (1854 onwards);

(c) Public General Statutes (1861 onwards);

(d) Current Law Statutes Annotated (1948 onwards); n.b. the annotations are not up-dated; and

(e) Current Law Statute/Legislation Citator (1947 onwards).

Halsbury's Statutes

(a) Search in the volume 'Tables of Statutes and General Index' using the alpha-betical, subject or chronological index;

cont.

(b) note volume number, page number and subject heading;
(c) check to see if volume has been re-issued after the index; if so, use volume index;
(d) find text;
(e) use volume number and page number in the cumulative supplement;
(f) use volume number and page number in the noter-up volume; and
(g) check *Is It In Force?*

5.5.2 Statutory instruments

Finding a statutory instrument

Citation

(a) By name: The Land Registration Rules 1925; or
(b) by number passed in year: SI 1978 No. 1846 (i.e., The 1846th Statutory Instrument laid in 1978).

Search by using a citation

(a) HMSO Official Series (arranged numerically);
(b) Halsbury's Statutory Instruments (arranged by subject); but note that not all are full text;
(c) www.open.gov.uk (since 1988);
(d) www.justis.com (since 1987);
(e) LEXIS or Westlaw.

Search by subject

(a) *Halsbury's Laws and Halsbury's Statutes;*
(b) *Current Law Legislation Citator;*
(c) *Halsbury's Statutory Instruments;*
(d) Texts and Practitioner Texts;
(e) LAWTEL;
(f) House of Commons Weekly Information Bulletin; or
(g) Current Law Yearbook and Current Law Monthly Digest;

Tracing an SI made under a Statute:

(a) Check the Table of Statutes in the Service binder to *Halsbury's Statutory Instruments* and cross-check with the *Chronological List of Instruments*
(b) *Current Law Legislation* Citator notes under each statute every SI made since 1947 referencing through to the *Current Law Yearbook.*
(c) *Halsbury's Statutes* notes SIs made under each Statute section
(d) *Current Law Monthly Digest.*

Tracing repeals, commencements and amendments:

(a) *Current Law*—the section entitled Dates of Commencement;
(b) Annual List of *Statutory Instruments;*
(c) *Halsbury's Statutory Instruments* notes amendments and repeals;
(d) In *Halsbury's Laws* and *Halsbury's Statutes* the cumulative supplements and noters-up indicate alterations and repeals;
(e) Table of Government Orders; or
(f) LEXIS—which, however, only gives current text as amended.

cont.

Halsbury's Statutory Instruments

(a) If you know year and number, use the Chronological Index in Binder 1 to find the title;
(b) use Chronological Index in relevant volume (or in monthly update directly);
(c) check in volume: SI will either be note, summary or full text; and
(d) Check Monthly Update.

If you know the title:

(a) Use the Alphabetical Index in Consolidated Index;
(b) then look in Chronological Index in relevant volume (or in monthly update directly);
(c) check in volume: SI will either be note, summary or full text; and
(d) check Monthly Update

5.5.3 Cases

Finding a case

Search by subject

(a) *The Digest* (includes Commonwealth and European cases);
(b) the law reports index (updated in All England Law Reports and Weekly Law Reports);
(c) *Current Law*;
(d) *Halsbury's Laws*;
(e) practitioner textbooks; and
(f) LAWTEL, Current Legal Information, Westlaw.

Search by Statute

(a) *Halsbury's Statutes*;
(b) *Current Law Legislation Citator*;
(c) Statutes judicially considered in law reports index.

Search by Statutory Instrument

(a) SIs judicially considered in law reports index; and
(b) *Current Law SI Citator in Current Law Legislation Citator* (since 1993 only)

Search by words and phrases

(a) Judicial dictionaries;
(b) words and phrases in judicially considered;
(c) words and phrases in law reports index;
(d) words and phrases in *Current Law*;
(e) words and phrases in *All England Law Reports index*;
(f) words and phrases in *Halsbury's Laws* index;
(g) words and phrases in the *Digest* index; and
(h) LEXIS or Westlaw.

Updating a case

(a) The Digest;
(b) *Current Law Case Citator* (references given to paragraphs in *Current Law Yearbook*); and
(c) table of cases judicially considered in law reports index.

5.6 Doing library-based research

It may seem strange to include a section on library-based research when you may feel that one skill you definitely have by this stage is the ability to do legal research. Certainly, your existing research skills are by no means irrelevant. However, they are unlikely to be enough. The jump from the academic to the practising environment involves the development of some rather different research skills, and the use of research resources with which you may have little or no familiarity.

The actual process of research can be broken down into four phases (see further Price, Bitner and Bysiewicz, *Effective Legal Research*—para. **5.14**):

(a) analysis of the problem;

(b) preliminary review of the subject-matter;

(c) search of primary and secondary sources;

(d) updating of search.

5.7 Analysis of the problem

The starting-point for research must inevitably be the client's problem. But clients' problems may be presented in very different ways.

During your training contract, research problems will be presented to you by your principal and, increasingly, by your own clients. Inevitably, therefore, the form in which these problems are presented will vary, but they will typically fall within a broad continuum, which goes something like:

- Can you find out when class 1A National Insurance contributions will be payable on a company car?
- I read an article a while ago which mentioned an important case on the jurisdiction of national courts to apply the competition rules under Art. 81(2) of the EC Treaty. I think the name was 'Auto' something or other. Can you find a law report for me?
- Mrs Smith called in to see me about the house we exchanged contracts on last week. Apparently it was damaged in the storm the other night. The vendor is now saying that responsibility for repairs has passed to her. Can you work out what the position is on that, please?
- I own a shop in the High Street. The cellar of the shop runs under the road. About six weeks ago, part of the cellar roof collapsed. I've had it shored up, but now part of the road surface over the cellar has started to sink and crack. Today I got a letter from the council informing me not only that immediate repairs to the cellar must be carried out, to their specification, but also that they intend to charge me for the cost of repairing the road surface. What should I do?

What is it that makes the last of these problems more difficult to research than the first? The answer should be obvious. In the first, the problem is discrete and already reasonably well defined; we know the basic subject-matter, and it is sufficiently narrow to enable us to find a solution quite easily—assuming we know where to look. The second problem is only marginally more difficult. In such cases, we can go straight to the library and commence a search using the subject indexes of the appropriate sources. There is little need for problem analysis. The third problem requires a little more problem analysis, but is still fairly well defined, though here our research strategy would need to be rather different because of the greater reliance on factual as opposed to legal criteria for defining the problem. In the last, the problem is unfocused. It is not particularly clear what the issues are, or how, or where, the answers will be found.

One of the key skills in practice is to identify and research unfocused problems quickly and efficiently. It is important because clients simply do not walk into the office with a neatly packaged problem labelled (for example) 'enforceability of covenants over land', or 'powers of highway authorities'. To be sure, the lack of focus will not always create difficulties. Your knowledge of the law may be sufficient for you to identify the problems and either give advice then and there, or to at least identify the issues sufficiently to make the process of research relatively straightforward. But this will not always be the case.

In determining what needs to be researched, you must first depend on the basic skills of analysis and classification already discussed. Your first job is to determine what your client wants advice on. Take another look at the last of our four research scenarios. The problem raises a number of potential issues: is your client in breach of any common law or statutory duty? If so, what? If so, can the council require your client to make immediate repairs? Can it legally charge the client for the damage to the road? Even if these rights exist, has the council used the proper enforcement procedures? If your client has employed contractors to shore up the roof, have they done the job adequately? Could there be a case of negligence here, with the possibility of an action or set-off against the contractors? And so on. Each of these issues may centre upon substantially different questions of fact, law or procedure. Before you commence research it is therefore essential to separate problems into their constituent parts: to identify the factual issues and both the substantive law and procedural issues which arise.

Your capacity to identify the issues will depend in part upon your existing knowledge of the field. If you already have a reasonable knowledge of the area, you will probably be familiar with the research resources you need to use. You may have already identified key primary sources that you might need to look at. But at least some of the problems you will be confronted with will involve what are, for you, novel issues. In this sort of situation, you need to develop a capacity for researching problems from scratch—a technique that is often referred to as 'cold starting' (see, e.g., Tunkel, *Legal Research*). Cold starting involves the use of keywords to formulate a research strategy. Keywords are terms which you have identified from your client's story which you consider to be significant in identifying and structuring the problem. These keywords may be based on either the specific facts of the case, or on any legal concepts which you have identified as relevant. For simplicity we shall call these techniques 'key-fact' and 'key-concept' searching.

5.7.1 Key-fact searching

If you have little idea of the law involved, you will be forced to rely on key-fact searching. The trick in picking out keywords from the facts of an issue is to try to select those terms which are most likely to provide useful keywords. A simple technique which helps in this is to look for 'PEPT words'.

The acronym PEPT stands for parties, events, places and things. Together these four groupings provide the categories of facts which, in the majority of cases, will yield useful keywords. This does not, of course, mean that each category will yield relevant terms in every situation. For example, it will not be significant in determining criminal liability for burglary that the person you are defending is unemployed—though that fact may be significant if the issue you are researching is one of sentencing. You need to use some common sense, and your knowledge of law, to determine broadly what words are most likely to produce results. Relevant points you should consider when using the PEPT categories are:

Parties

Do the persons involved belong to a particular class—independent contractors, minors, lessors or lessees, company directors etc? Are they in a significant relationship—e.g., as cohabitees, vendor and purchaser, doctor and patient, solicitor and client?

Events

What event(s) created the problem—e.g., selling shares, driving a car, conveying a flat, etc.?

Places

How important is the place (and/or time) where (when) the event happened? For example, did an accident take place on the street, at work (during or out of normal working hours), on private property, etc.?

Things

Does the case involve a specific object (a van, house, computer or 50 tonnes of sheet steel) which might be relevant to the legal issue?

Keywords discovered using PEPT can form the basis for any subsequent literature or case searches that you might wish to perform. Key-fact searching is a useful device in its own right, particularly if you are using your facts to try to identify a potential cause of action. However, in most cases you will probably have enough knowledge of the area to identify relevant legal concepts so that it is possible to integrate key-fact and key-concept searching to provide a more precise search strategy.

5.7.2 Key-concept searching

Key-concept searching involves identifying the legal concepts involved in the problem. In so doing, it is advisable first to distinguish between issues of substantive and procedural law.

In substantive law terms, keywords can be established by thinking about three things:

(a) the cause of action or crime charged;

(b) any defence available; and

(c) in civil cases, the remedy sought.

So, to give a simple example, in a civil action for personal injuries the cause of action would be negligence, for which the claimant would be seeking the remedy of damages; the facts might disclose some assumption of risk by the claimant, thereby raising a volenti defence. Thus each of the terms 'negligence' 'damages' and '*volenti non fit iniuria*' could be used for key-concept searching.

Key-concept searching can similarly be used to help research procedural issues. The main difference from researching substantive issues lies in the fact that we not only need to think about the cause of action, defences and remedies, but also about *context*—the stages of the process in which we are engaged. Does a problem arise before or after exchange of contracts, at trial or pre-trial, for example? So, if a client wanting a divorce from her husband is concerned that he might attempt to avoid a substantial settlement by transferring assets overseas, there are a number of research issues you could follow up. You might want to explore whether the court determining the settlement has the power to take assets into account if they are outside the jurisdiction; but it could be more important to try to prevent those assets leaving the country in the first place, so are there orders available that would do this? Thus, your key concepts could involve the obvious terms, such as 'divorce', 'family property' and 'financial relief' (which are so general as to cover both substantive and procedural issues), and the more specific procedural concepts such as 'emergency procedures' or 'injunctive relief', or, if you have sufficient grasp of the area already, 'freezing injunction'.

5.8 Review of the subject matter

The simplest way to begin researching a problem is to find a book about it. Books about law, as opposed to books of law, which contain just the legislation or case reports, are often referred to as 'secondary' sources, to distinguish them from the latter, which are 'primary' sources.

At this stage in your research you should be attempting to obtain an orientation to the problem. Depending on your knowledge of the area, you may be seeking to obtain an overview of the law (the principles, concepts and language used) as well as basic references to primary (and possibly other secondary) sources. The purposes of review reading may therefore be:

(a) to familiarise yourself with the legal principles involved;

(b) to enable you to refine and/or extend your keyword search;

(c) to obtain references to primary authority which will form the basis of a search; or

(d) to obtain references to other secondary sources which may form part of the review.

Academic treatments of the subject, including established student texts, may be an adequate starting point, but they do have some limitations.

First, they are often organised according to conceptual criteria which do not always relate easily to the kind of fact-centred problems which will confront you in practice. This means, for example, that their indexes tend to be structured around concepts as opposed to fact keywords, which may limit their utility.

Secondly, they may not be updated as frequently as key practitioner texts—especially the loose-leaf volumes. You may therefore need to do more updating than would be necessary had you gone to a practitioner text or encyclopaedia.

Lastly, they will not necessarily give sufficient (or any) weighting to practical or procedural issues which could be relevant in determining your strategy in a particular case.

As a result, it may make better sense to go straight for an encyclopaedia or other practitioner text at the outset. We will identify the general sources you are likely to use as a practitioner, and offer some comments on their strengths and weaknesses. Advice on particular search techniques for these different sources can be obtained from more specialist works on legal research, such as Tunkel, *Legal Research*; Clinch, *Using a Law Library*; or Dane and Thomas, *How to Use a Law Library* (see 'Further Reading' at **5.14**).

5.8.1 Legal encyclopaedias

These take a number of forms and serve a variety of functions. Perhaps the best known, and most widely available, are the three Halsbury works: *Halsbury's Laws of England*, *Halsbury's Statutes* and *Halsbury's Statutory Instruments*. The two legislative series will be discussed later in this chapter. The Halsbury volumes are not the only major encyclopaedic works worthy of some mention, and we shall briefly consider a few other important resources before discussing *Halsbury's Laws of England*.

5.8.1.1 Specialist encyclopaedias

Most major areas of legal practice have their own encyclopaedic work or practitioner text. Most are in loose-leaf format and many are multi-volume sets containing the full text of legislation as well as commentary.

5.8.1.2 Words and phrases

The growth of legislation has meant that questions of statutory meaning are increasingly the dominant issues of law coming before the higher courts, to the extent that there are a number of specific resources identifying accepted judicial usage. *Stroud's Judicial Dictionary of Words and Phrases* by J. S. James (London: Sweet & Maxwell) and *Words and Phrases Legally Defined* (London: Butterworths) are among the more commonly used.

5.8.1.3 *The Digest*

For researching points of case law, *The Digest* may sometimes be useful. It provides very brief case notes, organised alphabetically under quite detailed subject headings, on decisions from most common law jurisdictions outside the United States. It is not the easiest of encyclopaedias to use, and it can be misleading in some circumstances, but is particularly useful if you need to find older cases (which might not appear in *Current Law*, for example) or material from the Commonwealth.

5.8.1.4 *Halsbury's Laws*

One of the most valuable starting-points, particularly if you are researching a topic about which you have little or no existing knowledge, is *Halsbury's Laws of England*. The current edition (the 4th) was only completed in 1991, and some of its 56 volumes (several now sub-divided) have already been revised and reissued. The whole work is regularly updated by a loose-leaf current service volume. *Halsbury's Laws* covers all areas of English law, by summarising the present state of the law with references to the relevant case law and statutes. The quality of its coverage is generally good, but it is not exhaustive! Its other great advantage is the detailed level of indexing, which makes it particularly easy to use for key-fact searches. For example, assume that you are interested in problems related to arsenic poisoning contracted from foodstuffs, you could actually look up 'arsenic' and find:

ARSENIC
  control, **18**, 1111
  importation in food, **18**, 1181

Each of these references gives us an indication of the context in which arsenic is being discussed, followed by the number (in bold type) of the volume in which it appears, and then the paragraph—not page—number.

Apart from the two-volume Consolidated Index, searching *Halsbury's Laws of England* involves using three separate elements of the work. From the index you can go direct to the subject volume which is relevant and make a note of what appears there. Updating that text can be done using the cumulative supplement and current service volumes. The supplements are published annually and contain all updates to the main volume, up to their own date of compilation. It is therefore only necessary to use the latest supplement. The current service provides an update on everything that has happened since the publication of the last supplement, in the form of monthly reviews and a noter-up section.

5.8.2 Court practice books

Most of these contain a mixture of commentary, primary sources and precedents. They tend to be published annually, or at least regularly updated. The oldest established are the *White Book,* Archbold's *Criminal Pleading, Evidence and Practice* and *Stone's Justices' Manual.* More recent texts include *Blackstone's Criminal Practice* and *Blackstone's Civil Practice.*

5.9 Searching primary and secondary sources

Your preliminary review of the subject-matter should have given you some references to primary authority, either legislation or case law, and possibly some references to books or journal articles also. Your first decision, therefore, must be where to go from here. It is not easy to determine just how much research is enough, and no book can give you anything other than a rather glib answer to that problem. For what it is worth our glib answer is that you should always do sufficient to convince yourself that you have the right solution. Whether this means you must plough through copious cases or rely on a textbook summary and analysis is for you to determine in the context of each problem. For the litigator it is important to remember that the best authority for a point of law is always primary authority, particularly in the higher courts and the more legalistic of our tribunals (such as the Inland Revenue Special Commissioners and the Employment Appeal Tribunal), and this factor, together with your own self-confidence, may ultimately determine your approach.

Assuming you consider it necessary to return to primary sources, there are one or two useful principles that are worth bearing in mind.

5.9.1 Searching legislation

This section is concerned with identifying and finding sources of both primary (statutory) and secondary (delegated) legislation.

5.9.1.1 Citing legislation

Before you can find legislative material it is helpful to have a basic idea of the formalities for citing both primary and secondary legislation.

The rules for citing statutes depend upon the age of the legislation, with the critical cut-off date being the end of 1962.

Modern Acts (post-1962) are conventionally cited by their short title and year only, thus: Social Security Administration Act 1992. The short title is normally defined by a section of the Act itself. Each Act is also given a chapter number (abbreviated to 'c.'). It is rarely cited these days, but can, in some circumstances, assist you in finding the Act. The citation of an Act as '1988 c. 33', for example, could only refer to the Criminal Justice Act 1988.

Acts passed between 1215–1962 are formally identified by chapter number and *regnal* year (that is a number assigned to the year in which it was passed, counting it from the first year of the relevant monarch's reign). Hence, the Wills Act 1837 is properly cited as '7 Will IV & 1 Vict c. 26'. The modes of citation can vary slightly between different sources, so that the various King Williams are sometimes cited as 'Wm' rather than 'Will' and the James's commonly appear as 'Jac' (from the latin—*Jacobus*).

The earliest Acts (i.e., pre-1548) did not generally acquire an individual title but were known collectively by the name of the place where Parliament sat during that particular session. Thus the title 'Statute of Westminster' or 'Statute of Gloucester' refers to more than one Act, so that each separate Act can only be identified by its chapter number. Conventionally the regnal year is now also cited to aid identification.

Many of the later Acts in this period acquired, either formally or informally, short titles, and the citing of regnal years gradually declined after World War II, until the system was formally dropped from 1 January 1963. However, the regnal year may (in the case of older Acts, will) still be required to trace legislation published in all the main series of statutes in print at the time—i.e., *Statutes of the Realm* (to 1713), *Statutes at Large* (to 1869), *Public General Acts and Measures* (from 1831), and *Law Reports: Statutes* (from 1861). A list of regnal and corresponding calendar years is contained in the *Guide to Law Reports and Statutes* (Sweet & Maxwell, 1959). The correspondence can also be worked out from Part I of the *Chronological Table of Statutes* (HMSO, 1992), though, because the statutes are listed by year, this is a very cumbersome way of going about it.

Delegated legislation has its own rules of citation, though, thankfully, these are simpler. The main modern form of delegated legislation is the statutory instrument (abbreviated to SI). These replaced Statutory Rules and Orders (SR & O's) after 1947.

Statutory instruments are cited by year followed by the reference number given to each instrument, thus: 'SI 1993 No. 790' refers to the Legal Advice and Assistance (Amendment) Regulations 1993. Since some practitioner libraries file statutory instruments numerically, it may be helpful to be aware of the number, though as all instruments in force are listed in *Halsbury's Statutory Instruments*, this can usually be found quite easily.

Statutory Rules and Orders are cited in the form 'SR & O 1915 (No. 33)'.

5.9.1.2 Finding statutes

Unadorned versions of statutes are published by Her Majesty's Stationery Office and in the *Law Reports: Statutes* series. These are now made available online by HMSO as well. If you can get access to them, however, there are three series which are potentially of greater practical value.

Halsbury's Statutes is a partner series to *Halsbury's Laws of England*, containing an annotated version of all legislation presently in force, organised by subject. It is therefore the main resource for discovering whether legislation exists on any topic.

The fourth edition of *Halsbury's Statutes* was finally completed in 1991. The techniques are similar to those employed in respect of *Halsbury's Laws*. Begin with the general index, which gives you a wide range of keywords, each of which will refer you to a volume and paragraph within the main collection. If you are looking for a specific statute by title, you can take a short-cut by referring instead to the alphabetical list of statutes at the beginning of the Table of Statutes and General Index. Any revisions which post-date the publication of the relevant main volume will be found in the cumulative supplement and the noter-up. The text of very recent legislation will appear, alphabetically by subject, in one of the five binders of the *Current Statutes Service*; it can be found by reference to the alphabetical list of statutes at the front of the *Service*.

In *Current Law Statutes* Annotated, all Acts are published in a loose-leaf format soon after enactment. Those which the editors judge to be sufficiently important are annotated by someone who is a specialist in that area of law. The annotations are not part of the Act, and do not, of course, have any legal force, though they can still be helpful in explaining the background, scope and operation of the Act. The current loose-leaf volumes are supplemented each year by bound volumes of the statutes, printed in order, according to the chapter numbers of the Acts.

Statutes in Force should be a useful collection as it is published by the Stationery Office in both microfiche and loose-leaf formats and is therefore easily and quickly updated when new legislation comes into effect. Material within the collection is organised under subject headings, but these are not always well indexed, so *Statutes in Force* can be rather difficult to use unless you have a very good idea of what you are looking for. The series is not annotated by any commentary on the provisions.

Statutes can also be found online in LEXIS and Westlaw. However, in both cases each section of an act receives its own individual entry. The text is also presented in its updated and amended versions.

5.9.1.3 *Hansard* and the search for statutory meaning

Since the House of Lords decision in *Pepper* v *Hart* [1993] 1 All ER 42, the courts have been entitled to refer to the Official Reports of proceedings in Parliament (*Hansard*) in cases where:

(a) the construction of the legislation is ambiguous or unclear; and

(b) there is a statement by the minister or other promotor of the Bill which clearly identifies the mischief or legislative intent underlying the obscure or ambiguous words.

The case law since *Pepper* v *Hart* shows that, having been let off the leash, the courts are willing to make substantial use of *Hansard* and often in circumstances that are outside some of the restrictions just identified (see, e.g., Holland & Webb, *Learning Legal Rules*, 4th ed. (London: Blackstone Press, 1999), chapter 8 for a review of recent developments). For advisors, especially in fields like tax and employment, or wherever there is a substantial statutory basis to the law, *Pepper* v *Hart* is far from being an esoteric nuisance. It creates a professional obligation to treat *Hansard* as a principal research resource whenever questions of statutory interpretation arise.

Researching *Hansard* requires a set of specific techniques which vary slightly, depending on the recency of the legislation.

For established legislation, the best starting point is the *Current Law Statutes* version of the Act. This will give you all the Hansard references to debates in both Commons and Lords. The commentary may well include some direct quotes from *Hansard* too. In recent years it has become more common for references to the discussion in Standing Committee also to be included, though that cannot be guaranteed, especially with older Acts. An alternative source for that information is the Stationery Office's *Catalogue of Government Publications*, which is issued on a monthly basis.

Current Bills, or very recently enacted legislation can be followed through Parliament via the Current Law monthly parts, which contain a 'Progress of Bills' section. The House of Commons *Weekly Information Bulletin* or Lawtel (see **5.11.2**) are good alternative sources.

Once you have identified the relevant debates, etc., these can be followed-up through *Hansard* itself. This is bound in separate volumes for the Commons and Lords, by Parliamentary session, rather than calendar year. Each volume is separately indexed, and there are separate sessional indexes. Current developments can be tracked through the paper parts of *Weekly Hansard*. In addition to paper and microfiche versions, the Stationery Office publishes the text of Hansard on CD-ROM. House of Commons and House of Lords *Hansards* are published as separate discs on a sessional basis, commencing with the session 1988/89 for the Commons and 1992/93 for the Lords. The main *Hansard* Commons series (whether hard copy or CD-ROM does not include reports of Standing Committees. They are published as a separate series. Unfortunately they seem to be stocked by even fewer academic and major public libraries than the main series, despite the fact that the reports of Standing Committees often contain the most useful and most detailed comments on the Bill. *Hansard* debates are now published on the Internet—see **5.11.4**, below.

5.9.1.4 Finding secondary legislation

We are less well served when it comes to secondary legislation. Her Majesty's Stationery Office issues all statutory instruments (or Statutory Rules and Orders as they were termed before 1948) individually and in annual bound volumes. These are now all available online as well at the HMSO website. Again, few libraries have the resources to maintain a full collection of these. The only other substantial paper source is *Halsbury's Statutory Instruments*, but this is by no means complete, and contains only a selection of the regulations in force, chosen on the basis of their perceived importance to the practitioner, and their unavailability elsewhere in the library. As with the other Halsbury series, the set is organised by subject-matter and the bound volumes are supplemented by a looseleaf service. This is one particular area where electronic information retrieval can be very helpful. Both Lexis and Westlaw carry full text versions of modern secondary legislation (see **5.11**, below). A number of practitioner texts carry the key SIs in their subject, sometimes in an annotated form (see, e.g., *Butterworths Family Law Service*; *Harvey on Industrial Relations and Employment Law* and J. Mesher, *CPAG's Income-Related Benefits: The Legislation*).

Local authority by-laws, which can be of considerable importance in some contexts, such as planning law, can only be obtained direct from the promulgating authority, for a fee.

5.9.1.5 Legislation of the Scottish and Welsh Assemblies

Under the recent devolution Acts, the Scottish Parliament and the Welsh Assembly each have the power to create legislation in their own right. The Welsh Assembly may only make secondary legislation in the form of statutory instruments (see s. 66(2) Government of Wales Act 1998). The Scottish Parliament, however, can create primary legislation, to be known as 'Acts of the Scottish Parliament' (s. 28(1) Scotland Act 1998). Both sets of legislation are published independently of the laws passed by the Westminster Parliament.

5.9.2 Searching case law

As you are probably all too well aware, case law research is complicated by the proliferation of generalist and specialist law reports. So, the first problem is determining where to look.

5.9.2.1 The range of law reports

In essence the law reports can be divided into two historical phases, pre- and post-1865.

Before 1865, there were two main sources of case reports:

Year Books

Established about 1285, these continued in existence until 1535. They were probably derived from the notes of cases taken by student advocates. They are not consistently reliable and of little more than academic interest these days.

Private Reports

Sometimes called 'Nominate Reports' because each series is named after the counsel who compiled them. Hundreds of reports appeared in this era. The same case could be reported by a range of reporters, and in such cases it is not unusual for the contents, even the decision, to vary as between reports! These reports are now collected together in *The English Reports*, which require a fairly specialised search technique of their own. Cases from the nominate reports are still cited, though increasingly rarely. Some of the most important cases from between the years 1558 and 1935 have also been reprinted in a series called the *All England Law Reports* Reprint (cited as All ER Rep.).

For case law post-1865, there are three further categories of case report:

The Law Reports and alternatives

Published by the Incorporated Council of Law Reporting, the so-called 'Law Reports' consist of four series, which since 1891 have been entitled: Queen's (or King's) Bench Reports (abbreviated to QB or KB), Chancery (Ch), Family (Fam)—since 1972, previously the Probate Division (P)—and Appeal Cases (AC). This broadly reflects the division of work between the superior courts, as Queen's Bench, Chancery, and Family Reports will contain reports of both High Court and Court of Appeal decisions. Appeal Cases contain House of Lords (English, Irish and Scottish) appeals and Privy Council cases.

The Council also publishes the Weekly Law Reports (WLR), which are relied upon by many practitioners. This is a generalist series of reports, covering cases in the superior courts, and some decisions of the European Court of Justice. The WLR are published in three volumes. Volume one contains those (less significant) cases which are not intended for republication in the Law Reports.

The All England Law Reports (All ER), published by Butterworths, are probably the most widely used alternative to the Council's publications. Like the WLR, they are also published weekly, and cumulated into usually three, sometimes four, annual volumes. The volumes here are organised chronologically, rather than by importance of the case. Inevitably there is considerable overlap between the contents of the All ER and WLR, but the order of publication of cases can sometimes differ quite markedly between each series.

The Times, The Guardian, the *Daily Telegraph*, and *The Independent* newspapers all publish case reports within their papers. *The Times* and *The Independent* report cases on a daily basis, but *The Guardian* and *Daily Telegraph* publish reports much less frequently, and tend to focus on public law issues, or cases that are otherwise newsworthy. Although *The Times* reports are the longest established and best known, all these newspapers' reports are capable of citation. After a gap of many years, *The Times* has also recommenced publishing its law reports separately in looseleaf format. Cases published in all these newspapers are indexed in a volume called the *Daily Law Reports Index*.

The Specialist Law Reports

There are now many series of law reports which are of considerable value to the subject specialist, and will often contain cases not reported in the generalist series. These include:

- Criminal Appeal Reports (Cr App Rep).
- Estates Gazette Law Reports (EGLR).
- Family Law Reports (FLR).
- Fleet Street Reports (FSR).
- Housing Law Reports (HLR).
- Industrial Tribunal Reports (ITR).
- Industrial Relations Law Reports (IRLR).
- Justice of the Peace Reports (JP).
- Lloyd's Law Reports (Lloyd's Rep).
- Pension Law Reports (PLR).
- Property, Planning & Compensation Reports (P & CR).
- Road Traffic Reports (RTR).
- Simon's Tax Cases (STC).
- Tax Cases (TC).

Unreported cases

There are still many cases, even in the higher courts, which go unreported. This does not mean that they are wholly irrelevant. Transcripts of unreported Court of Appeal cases are available to judges and practising lawyers from the Supreme Court Library in London; transcripts from the Court of Appeal and the Crown Office list (among others) are also available on electronic databases such as Lawtel, LEXIS and Westlaw. Unreported cases can quite properly be cited to a court, provided they establish a principle which could not have been established through reported case law—see per Lord Diplock in *Roberts Petroleum Ltd v Bernard Kenney Ltd* [1983] AC 192; [1983] 1 All ER 564; also *Practice Note* (*Court of Appeal: authorities*) [1996] 3 All ER 382.

5.9.2.2 Finding cases

There are a variety of indexes you can use to track down cases. *Current Law*, which is issued in monthly parts and then consolidated into the *Current Law Year Books*, is indexed by case name and subject. The *Law Reports* series also publishes cumulative indexes which are organised by case name and subject. These contain not only references to the *Law Reports*, but also to the *Weekly Law Reports*, the *All England Law Reports* and a number of specialist series as well (such as the *Industrial Cases Reports*).

If you are engaged in litigation, particularly before the higher courts, then, strictly speaking you should bear in mind that the *Law Reports* series (Appeal Cases, Chancery, Family and Queen's Bench) take precedence of citation. It makes sense in such situations to use those (if possible) from the outset, as it could save rechecking references later.

Decisions of the Courts of Appeal and the House of Lords are now made available online at the Lord Chancellor's website. Cases are also available in transcript from Smith Bemal and via Casebase and Casetrack databases.

5.9.3 Searching EC law

The complexity and volume of Community law makes EC-focused research an increasingly specialised concern. In this section, we shall first outline the primary sources of EC law before discussing basic research methods.

5.9.3.1 The sources of EC law

EC Law is conventionally divided into primary and secondary sources. This distinction is critical, since very different research questions and strategies arise in respect of each form of legislation.

The *primary sources* comprise the various treaties, which include:

(a) the founding treaties of the three communities underpinning the EC—the Treaty of Paris 1951 (European Coal and Steel Community); the Treaty of Rome 1957 (EURATOM) and the Treaty of Rome, or EC Treaty, also of 1957.

(b) the amending treaties, including the Single European Act 1986 and the Treaty on European Union (TEU) 1992, and the Treaty of Amsterdam (ToA) 1997, which, once ratified, has introduced a number of new measures streamlining decision-making in the EU, and otherwise amending both the TEU and EC Treaty. As part of this process the Amsterdam Treaty has renumbered all the Articles contained in both the EC Treaty and TEU; in this section we have indicated the new numbers in square brackets following the original derivations.

(c) the accession treaties which have allowed new member States to enter the Community.

(d) subsidiary treaties that may also be regarded as a source of EC law, either where agreements have been concluded by the EC itself (e.g., under Art. 113 EC Treaty [Art. 133]) or, more exceptionally, by member states individually, under Art. 220 EC Treaty [Art. 293] (e.g., the 1968 Convention on Jurisdiction and Enforcement of Judgments in Civil and Commercial Matters).

The *secondary sources* comprise legal acts of the European Commission, or Council of Ministers, as follows:

(a) Regulations, which have direct application. Regulations apply to all member States and individuals, and are binding without further action on the part of the member States. Regulations have to be published in the *Official Journal* of the Community. Regulations are cited in the form: 'Reg. (EEC) 1408/71'—note that the source of the regulation (i.e., the treaty under which it is promulgated) is formally cited in brackets, followed by the number of the regulation (1408) and the year (here 1971).

(b) Directives also have general application in that they are binding on member States as to the *result* to be achieved, but leave open to each State the *form and method* of implementation. Member States are invariably given a time period for implementation. A Directive has no direct effect in law until it is implemented by the State (unless the time period has passed), and the method of implementation will vary from State to State according to their own special conditions. Directives are cited in the following form 'Dir. 85/374/EEC' indicating the year promulgated; the number of the Directive and its legal source. The order is thus reversed from that used in citing Regulations.

(c) Decisions are not something of judicial origin, but binding 'orders' issued by an institution of the Community and addressed to an individual or State. These frequently arise in competition cases where the Commission can be asked to determine the legality of agreements. Decisions are cited in the same form as Directives.

(d) To these three binding legal provisions we can also add Recommendations and Opinions which are acts of the Commission or Council which are of persuasive authority only. These are also cited in the same way as Directives.

5.9.3.2 Finding primary sources

The primary sources of Community law remain generally less easy to access than their English counterparts.

For the practitioner, the main sources for the Treaties therefore tend to be Vol. 50 of *Halsbury's Statutes* or Sweet & Maxwell's *Encyclopedia of European Community Law*, 'B' volumes. These are perfectly adequate for most purposes.

5.9.3.3 Finding secondary sources

EC secondary legislation (i.e., Regulations, Directives and Decisions) is more of a problem. All secondary legislation is recorded, from draft to final stages, in the EC's *Official Journal*. The great advantage of the *Journal* is its speed of publication. It is by far the most up-to-date source on Community legal and fiscal issues. This strength is partly offset by the search strategy required to use it, which is nearly as complicated as some of the EC's legislation (for concise explanations of the strategy, see Holland and Webb, *Learning Legal Rules*, or Tunkel, *Legal Research*). As it is, few practitioners will have direct access to the *Official Journal* (unless they are close to a European Documentation Centre or other university law library). There are essentially two commercially published alternatives: Sweet & Maxwell's Encyclopedia of *European Community Law*, and CCH's *Common Market Reporter*, both of which are multi-volume loose-leaf works.

In research terms, there are two particular problems you need to be aware of when dealing with secondary legislation.

First, secondary legislation, especially Regulations, can be quite frequently and extensively amended. Therefore it is critical to ensure your version of the text is up-to-date. Updates are published for both the *Encyclopedia* and the *Common Market Reporter*, but, as with most loose-leaf works, there is usually some time lapse involved. So, another useful source for checking the current status of legislation is Butterworth's *European Communities Legislation: Current Status* (1992). This is published in two volumes, with cumulative updates published in May, August and November each year. The volumes are split chronologically, with the first volume containing legislation up to and including 1980, the second running from 1981 to date. Each act is listed by its legislative act number. There is a short description, an indication of whether it is in force or repealed and, if in force, there is a list of any amendments and/or partial repeals made. If you do not have access to this, *Current Law* and now *European Current Law* also contain brief details of important changes in EC law.

Second, the principle of direct effect means that it is extremely important to know whether the UK has passed legislation to implement an EC Directive or not. There are two means of checking this.

One is to use the *Encyclopedia of European Community Law*. At the beginning of volume 'A1' ('United Kingdom Sources') there is a table of selected Community Secondary Legislation, which cross references the EC legislation with the implementing UK legislation which appears in volume 'A' of the *Encyclopedia*. Do remember, however, that the coverage of the *Encyclopedia* is not complete, though it is adequate for most purposes.

The other is to use *Butterworth's EC Legislation Implementator.* This is published twice yearly and lists all Directives which have been implemented by domestic legislation, in chronological order. Unlike the *Encyclopedia*, the *Implementator* does not contain any legislative texts at all.

5.9.3.4 EC case law

Case law in the EC is made by the courts of the Communities and by national courts. There have been, since 1989, two EC courts: The Court of Justice of the European Communities (CJEC or, more commonly abbreviated to, ECJ) and the Court of First Instance.

The ECJ has a wide jurisdiction given to it under the Treaties. Here are a few examples:

(a) It exercises judicial control over the institutions of the Community such as the Commission.

(b) It exercises powers of judicial review on the validity of Community legislation such as Regulations and Directives—in other words it can annul Community secondary legislation.

(c) It hears cases brought by member States or Community institutions against other member States regarding violations of the Treaties.

(d) It decides whether international agreements entered into by the Community are compatible with the provisions of the Treaties.

(e) It gives preliminary rulings under Art. 177 [Art. 234] when asked to do so by a court of a member State concerning the interpretation of Community legislation.

Since 1989, it has also acted as an appeal court to a new court, the Court of First Instance (CFI), which deals with, amongst other things, competition cases, cases relating to the staff of the Community and certain cases under the ECSC Treaty.

Cases before the European courts are cited by reference to case names (in the English style) and number. The case number is an important feature of decisions. The proper mode of citation is to give the case number first, followed by its name, and then the citation of any report. Thus, for example, one of the leading cases on the impact of EC membership on national sovereignty, is properly cited as case 6/64 *Costa v Ente Nazionale per l'Energia Elettrica (ENEL)* [1964] ECR 585. The case number, it can be seen, contains two elements: the actual number at which the case was listed (here, No. 6) and the year in which the application or reference was made (1964)—this is not necessarily the year in which the case was heard.

Since the creation of the Court of First Instance, a new element has been introduced into the citation of cases. Now all cases have the prefix 'C-' or 'T-' before the number. This indicates that the case was listed, respectively, either before the ECJ or the CFI, thus case C-229/89 *European Commission* v *Belgium*.

The national courts of member States are an increasingly important source of case law on EC matters. In terms of precedent, the English courts will treat their own judgments as being binding in the normal way, subject always to the power of a court to make a reference to the ECJ under Art. 177 [Art. 234] EC Treaty. Decisions of domestic courts in other member States may be regarded as persuasive authority on EC law.

This has had some impact on the research strategies necessary when dealing with EC matters. It does mean that you need to spread your net fairly widely. Decisions of English courts appertaining to EC law will be reported in the usual English law reports, and/or in the *Common Market Law Reports* (CMLR), which are an unofficial series published monthly, in English. Domestic cases from other member States will also appear in the CMLR, but so far have not tended to appear in the English law reports.

The case law of the EC court is best accessed by reference to the *Common Market Law Reports*, which are published far more quickly than the official *European Court Reports*. Important ECJ cases are now starting to appear more frequently in both generalist and specialist English law reports also.

Commentaries on the case law and legislation appear in a wide variety of generalist and specialist encyclopedias, including *Halsbury's Laws of England* (Vols 51 and 52), as well as more specialist textbooks.

5.9.4 Materials on the European Convention on Human Rights

The partial incorporation into English Law of the European Convention on Human Rights (ECHR) is likely to have a significant impact on the range of research required, particularly in the areas of practice most susceptible to challenge under the Human Rights Act 1998 (HRA), such as criminal litigation, public child care law and education.

While the provisions of the ECHR which are incorporated into English Law as 'Convention rights' are contained in Sch. 1 HRA and therefore do not need to be tracked back to the Convention, the same cannot be said of the extensive caselaw of the European Court of Human Rights (ECtHR). This remains an important potential legal source, notably under s. 2(1) HRA, which requires UK courts to take the jurisprudence of the European Court into account when interpreting Convention rights.

The most commonly accessible English-language version of European Court decisions is the *European Human Rights Reports* (EHRR), published by Sweet & Maxwell. However, decisions can also be accessed in full-text version from the Council of Europe's website though not all decisions there are published in English. Selected decisions of the ECtHR also appear in English in the *International Law Reports*, published by Grotius, Cambridge.

Among the more practitioner-friendly texts on the Convention and/or the HRA are Starmer, *European Human Rights Law* (Legal Action Group, 1999) and Wadham & Mountfield, *Blackstone's Guide to the Human Rights Act 1998* (Blackstone Press, 1999). There is one specialist journal dedicated to the field, the *European Human Rights Law Review*, though there has also been substantial coverage of incorporation issues in the generalist journals, particularly *Public Law*. *Legal Action* has also run a number of short articles on the Convention/HRA as well as carrying regular reviews of decisions of the ECtHR and European Commission (though note that the latter's function is now replaced by a preliminary hearing procedure before a panel of the ECtHR—see the 11th Protocol to the ECHR).

5.9.5 Searching for journal articles

If you wish to find an article on a particular subject there are three paper sources to use.

The *Legal Journals Index*, which commenced publishing in 1986, contains full details of a very wide range of legal journals published in the United Kingdom, and so provides an extremely valuable research resource. It is indexed according to both subject-matter (with a brief summary) and name of author. Cases and Acts of Parliament which have been the subject of a commentary are also indexed under their title. *Legal Journals Index* is now available through Westlaw.

Current Law lists recent articles under each subject heading, and each *Current Law Yearbook* contains a separate index thereto. The range of journals covered is not exhaustive and so it is less useful than the *Legal Journals Index*. It does, however, provide some assistance in tracing articles before 1986.

Very exceptionally it may be helpful to look further afield, in which case there is the *Index to Legal Periodicals* and the *Index to Foreign Legal Periodicals*. These are both American publications. The former is an index to all American journals, plus a selection from Britain, the Republic of Ireland and the Commonwealth; the latter indexes articles on international and comparative law, and on the municipal law of all countries which do not appear in the *Index to Legal Periodicals*. They tend to be available only in the larger and more specialist libraries.

In addition to these paper resources, there are a number of electronic resources which might prove useful. Of particular value are the Lawtel *Articles Index* and Sweet & Maxwell's *Current Legal Information* service—which contains, *inter alia*, an electronic version of the *Legal Journals Index*. Among the more esoteric resources (from a practitioner perspective), Lexis holds an extensive range of US journals in full text form, and specialist Internet sites like Findlaw.com provide access to electronic law journals based on the Internet.

5.10 Updating the search

Publication time lags mean that most research will need occasional updating to check its accuracy. Updating techniques will be considered under the same headings as the initial search strategies.

5.10.1 Updating on legislation

For general updating purposes, there are a number of sources that can be used. As we have already seen, these include the noter-up features of the Halsbury works, and other loose-leaf sources. *Current Law* is also a useful general source in finding out what has happened, as it contains a brief description, under the relevant subject heading, of any legislation passed during the year. It is not sufficiently detailed to be of substantive help, but it can make you aware of primary or secondary legislation which you might not otherwise have known about.

When updating legislation, however, we tend to be concerned with two specific questions:

(a) Is the legislation in force?

(b) Has the legislation been amended or repealed?

In determining whether legislation is in force there are several sources to be used if you are to be completely up to date. Your usual starting-point should be a copy of *Is It in Force*? This is part of *Halsbury's Statutes* and is published annually covering statutes passed in the last 25 years. The implementation (or non-implementation) of every section is listed there, alphabetically by the year and short title of the Act. To complete the update it is necessary to go to either *Halsbury's Laws of England*, where commencement orders are listed in the noter-up, or to the *Current Law* monthly parts, where commencement orders appear under the relevant subject heading. This will bring you up to date to within a matter of weeks. Any final checking beyond that becomes more difficult. Her Majesty's Stationery Office publishes a *Daily List* incorporating such information, but this is not carried by all libraries. Otherwise it is a matter of checking the pages of journals such as the *New Law Journal* or *Law Society's Gazette*. If all else fails, a phone call to the House of Commons Information Service may succeed, provided your inquiry is sufficiently specific.

If you wish to discover whether legislation has been amended or repealed, the quickest technique is to use *Halsbury's Statutes* if you are looking for detail on amendments or partial repeals, or the *Chronological Table of Statutes* if you just wish to know when an Act was wholly or partly repealed or amended. The *Chronological Table of Statutes* is published in two volumes by Her Majesty's Stationery Office. It contains details on all Acts passed since the Statute of Merton in 1235. Its one substantial weakness is that it is published every other year, so that it is always at least a year behind and cannot be relied on for totally up-to-date information. The *Current Law Legislation Citators* may also be used in respect of post-1947 changes, but they do not give the full legislative history of Acts passed before 1947. The *Citators* contain information on all legislation passed in 1947–71, 1972–88, 1989–95 and 1996–2001 respectively. Each citator is organised chronologically by year and Chapter number (it also gives the title). Any changes to an Act are then listed section by section. This is important; it means that to use the citator most effectively you need to know not only the short title and chapter number, but also the specific sections of the Act that you wish to trace.

Suppose, for example, you wish to discover what has recently happened to s. 1 of the Criminal Evidence Act 1898 (c. 36). The latest changes (if any) will be contained in the citators, and subsequent *Current Law Year Books* or monthly parts. The 1972–88 citator has the following entry:

s. 1, amended: 1979, c. 16, s. 1; repealed in p. 1982, c. 48, sch. 16; 1984, c. 60, s. 80, sch. 7.

The 1989–95 Citator also contains an entry, thus:

s. 1, amended: 1994, c. 33, s. 31, sch. 10; repealed in pt.: *ibid.*, schs 10, 11.

These show that part of the section is no longer in force and that other elements of the wording have been amended. To work out the detailed effect of those amendments you would, of course, have to compare the various texts. The titles of those later Acts can either be found elsewhere in the citator, or you could go direct to the relevant volumes of *Halsbury's Statute*, or *Current Law Statutes Annotated*. More recent checks can be made using the statute citator in the monthly parts of *Current Law*.

Updating delegated legislation is best undertaken using *Halsbury's Statutory Instruments*, though Current Law provides a possible alternative to this. *Halsbury's Statutory Instruments*, like the Statutes, contains both an Annual Supplement and Monthly Survey for the current year. The Monthly Survey gives a brief summary of the effect of new statutory instruments, but it will usually be necessary to get hold of the full text to work out the effect of detailed changes. More recent updating can be done through either specialist intelligence services, or the Stationery Office *Daily List*.

Halsburys also publishes a comprehensive citator.

Using Westlaw and LEXIS will help as the text of the statute is presented in the form currently amended. Westlaw—through the analysis tab—gives detailed information on commencements.

5.10.2 Updating case law

The best method of updating case law involves finding the citation in the *Current Law Case Citators*. There are three bound volumes—the first covers cases reported from 1947–76, the second from 1977–88 and the third from 1989–95—a fourth soft-cover volume covers 1996–99. All reported cases are listed in the citator in alphabetical order, based upon the first-named party to the case. The citator will then give a list of all citations of that case, and a reference to a summary of the case in the relevant annual volume of *Current Law Yearbook*.

In addition to the citations of the case named, the citator also charts its subsequent history. If a case predates 1977, it is still worth checking the later citators, as they will tell you when the case has been most recently considered. If it appears, it will still be necessary to go back to the earlier citator, as the whole history will not be given in the latest entry. The form of reference in the citators is broadly the same so, to give an example, in the 1989–95 citator you will find (approximately) the following entry:

R v Secretary of State for Transport, *ex parte* Factortame [1990] 2 AC 85; [1989] 2 WLR 997; [1989] 2 All ER 692; [1989] 3 CMLR 1; [1989] COD 531; (1989) 139 New LJ 715, HL; reversing (1989) 133 SJ 724; [1989] 2 CMLR 353; (1989) 139 New LJ 540, CA ....*Digested*, 89/**3081**:

Considered, 90/9, 2125; 91/70; 93/1348; 94/3046
Distinguished, 92/1489

This gives us quite a lot of information about the *Factortame* case. In addition to the references to the various reports, we know that it appears in the *Current Law Yearbook* of 1989 at para 3081 and that it has been considered judicially in five later cases (two in 1990, one in 1991 and one each in 1993 and 1994), which are also digested in *Current Law*. Note that where a court has done more than merely consider a case, the citator will say so, using the conventional terminology of approving, distinguishing, overruling etc., so we can begin to assess the importance of those later decisions directly from the citator.

For very recent cases, it may be sufficient to use the monthly parts of *Current Law*, by searching through either the abbreviated case citator, to be found near the centre of each monthly issue (if you know the case name), or the relevant subject headings. However, there is also a publication called the *Daily Law Reports Index*, which, since 1986, has carried digested notes of all cases reported in the broadsheet newspaper law reports. Cases can be accessed by subject matter, case name, or the name of any legislation referred to in the case. The *Daily Law Reports Index* is normally up-to-date to two weeks preceding publication.

5.10.3 Updating EC law

Probably the most up-to-date source after the *Official Journal* is Butterworths *EC Brief*, which is published as a weekly newsletter covering the whole range of EC legal activities, together with a telephone support service. A number of the large law firms also publish their own briefing papers on EC matters. The coverage in these tends to be far more selective and in the form of news items or short articles, rather than the more basic legal intelligence of the Butterworths publication.

For some years, *Current Law* contained an EC law section. However, since the beginning of 1992 this has been replaced by a separate publication entitled *European Current Law*, which covers every country in Europe, except Turkey. The format is essentially the same as for *Current Law*.

More generally, the noter-up to *Halsbury's Laws* will also provide you with a general update of the issues covered by Vols 51 and 52 in the work.

For information on proposed legislation and other policy developments, there are a number of possible sources. The *Official Journal* will carry drafts of legislation when the Commission is seeking the opinion of the European Parliament, or the Economic and Social Committee. Draft Commission legislation will also appear in a bi-monthly

Community publication called *European Access*, usually in advance of any publication in the OJ. The European Commission also publishes briefing papers and policy documents in a number of areas, such as consumer protection, competition law and social policy. These can be obtained by contacting the Commission's London offices, and being placed on their mailing list. Some documents are provided free of charge, but the more substantial (and more useful) normally have to be paid for.

5.11 Using electronic information retrieval

There has been a mass explosion of electronic legal information retrieval in solicitors' offices.

Anyone intending to use a database system is advised to get trained on the system first. There is no substitute for that hands-on experience. Accordingly we will not attempt to teach you how to use the systems here but merely outline their uses.

Electronic databases are certainly efficient tools for updating cases or legislation, or doing specific 'words and phrases' definitional searches. Because of the way the search commands are implemented, however, they are less effective at more conceptual searching, and the onus is very much on the user to find an appropriate research strategy, as there tends to be little conventional indexing of the databases. Useful guides to search technique are published by the system suppliers.

5.11.1 LEXIS

The LEXIS database is one of the most extensive. It includes full text case reports of most cases reported in England and Wales since 1945 (tax cases since 1875); all public general Acts and statutory instruments in force; a small range of English periodicals; certain Scottish, Irish, Australian and New Zealand law reports; decisions of the Court of Justice of the EC; some French cases and legislation (in French) and extensive American materials. The volume of information carried by the system does mean that the system is not updated as rapidly as some practitioners might like, with delays of up to two months on some parts of the system.

5.11.2 LAWTEL

The LAWTEL database contains a wide range of material, much of which is not available on LEXIS, including the progress of Bills through Parliament, decisions of a number of specialist tribunals, and white and green papers. This is in addition to the 'normal' fare of leading cases, and primary and secondary legislation. The limitation of the system is that the material is mostly summarized, rather than full text, which rather restricts LAWTEL to an updating or initial research function. However, it links through to government websites for statutory material. The system, however, is generally well indexed, cross-referenced and quite easy to use. It has a daily update feature.

5.11.3 Westlaw

Westlaw is also an extensive database covering cases, statutes and statutory instruments and journal articles. The text of the statutes and statutory instruments is presented in its current (amended) form; although Westlaw does have a historic statutes section.

Westlaw holds full text case reports from the Law Reports plus full text from a number of specialized series; its case locator section holds summaries of many other cases.

Westlaw is accessible by a simplified search screen which enables searching to be done by topic or by citation.

Westlaw also has an excellent internal citator and referencer through use of the 'analysis' tabs.

Westlaw is a very user-friendly database.

5.11.4 Law on the Internet

An increasing amount of primary legal data is accessible via the World Wide Web (www), including legislation, case law and international legal materials. Standard legal forms are also being made available across the Internet by bodies such as the Legal Aid Board, the Land Registry and the Lord Chancellor's Department. Though sometimes of variable utility, much of this information is freely accessible (i.e., not as a subscription service), as the following table of free websites shows.

Useful 'free' websites for legal research

Legislation:	www.opsi.gov.uk/legislation/ about_legislation.htm	Full text of Public Acts of the UK Parliament from 1988. Acts of the Scottish Parliament and Northern Irish Assembly (1999–). UK statutory instruments from 1987 and SIs of the Welsh Assembly (1999–).
Case law:	www.bailii.org	Reported decisions of the UK and English superior courts from various start dates after 1995. Also decisions of the Scottish Court of Session and High Court (1998–) and of the Northern Ireland Court of Appeal (1998–). European Court of Justice decisions since 1954 and all CFI decisions, plus a range of reported tribunal decisions.
	www.parliament.uk/ judicial_work/ judicial_work.htm	Full text versions of all House of Lords judgments since November 1996.
	www.lawreports.co.uk	Incorporated Council of Law Reporting website. Includes electronic case summaries published as W L R (D), updated daily.
	www.curia.eu.int	European Court of Justice site, including full text of judgments, weekly case list and official texts relating to the Court.
	www.echr.coe.int	European Court of Human Rights: searchable full text database of cases. Also has full text of the Rules of the Court and the ECHR and its Protocols.
UK Parliament:	www.parliament.uk	Gateway to House of Lords and House of Commons websites, with information on their functions and activities and access to the Information Offices of both Houses. Can also be used to access parliamentary publications.

Cont.

	www.parliament.uk/hansard/hansard.cfm	Full text of the House of Lords Hansard (debates) from 10.6.96 and full text of House of Commons Daily Hansard (debates); complete from the start of session 1997/98.
Newspapers:	www.telegraph.co.uk	*Daily Telegraph*
	www.ft.com	*Financial Times*
	www.newsunlimited.co.uk	*The Guardian*
	www.independent.co.uk	*The Independent*
	www.the-times.co.uk	*The Times*
Other resources:	www.open.gov.uk	Gateway to a vast range of sites created by central government departments and agencies (including the Lord Chancellor's Department, the Home Office and the Land Registry) and local authorities.
	www.smlawpub.co.uk www.butterworths.co.uk	In addition to subscriber online services, both sites include a number of free services.
	www.findlaw.com www.ilrg.com	US sites providing direct access to all electronic law journals on the Net (mostly American publications; mostly, but not exclusively, on IT and 'Cyberlaw' issues) and other services, e.g., access to specialist search engines, such as Law Crawler.

Apart from the free sites there are a number of subscription Internet services useful to lawyers.

5.12 Reporting the results of research

In practice it is important that you keep a written record of research you have undertaken. As a trainee, it is essential that your principal has some form of notes or memorandum that can clearly be understood and, if necessary, checked. Even once you are fully qualified, it is good practice to retain a written record of your research, particularly if anyone else is likely to need access to your research, but more generally because it enables you both to update and to double-check your research quickly and easily.

5.12.1 Criteria for recording research

There are a number of ways in which research can be recorded, and ultimately it does not much matter what system you use, provided it meets certain basic criteria:

(a) Your report should contain a concise summary of your research findings, with advice where appropriate. (It is usually helpful if you organise the notes so that the research summary and advice are clearly separated.)

(b) Your notes should make it clear what research resources you have used. All references should be given in sufficient detail to enable another practitioner to find them. This means, wherever possible, standard citations and abbreviation should be used.

(c) The date(s) of the search and the date(s) of publication of your source(s) should all be identified.

(d) Wherever possible, the record should relate to the facts of the case and your client's needs and objectives (if known). Research that is too abstract and too academic is a waste of your time and of no benefit to your client.

A method that is well established in North America, and becoming more common here, is to use a standardised research memorandum, often in the form of a legal research checklist. A sample checklist is set out below. It has been designed so that it can be used, either as it is or in an amended form, as a standard research proforma.

5.12.2 A legal research checklist

Background information
Researcher ....
Memorandum for ....
Commenced ...... Completed ....

Strategy
Material facts ....
....
....
Legal issue ....
....
Keywords:
Fact ....
Concept ....

Database search? YES/NO

Encyclopaedias and digests
... *Halsbury's Laws*
... *Halsbury's Statutes*
... *Halsbury's Statutory Instruments*

... *The Digest*
... Other loose-leaf
....
.... [name]
.... *Words and Phrases*

Checked:
... Cumulative supplement
... Noter-up
Updated to: ....

Checked:
... Supplement
... Service
.... [dates]

Other secondary sources
Books:
Name .... Date published ......
Name .... Date published ......
Name .... Date published ......
Name .... Date published ......
Journals:
Legal Journals Index checked? YES/NO
Current Law checked? ... yearbook ... monthly parts

Primary sources
Legislation
... Stationery Office copy
... *Current Law Statutes*
... *Statutes in Force*
... *Halsbury's Statutes*
... Other source
....
.... [name]
... SI/SR & O

Case law
... *Law Reports*/WLR etc.
... Specialist series
... Daily law reports
... Internet

Cont.

Updated by:	
... *Is It In Force?*	... Case citator
... Statute citator	... *Current Law* (monthly)
... *Current Law monthly parts*	... *Law Reports Index*
... *Halsbury's Statutes* (noter-up)	... *Daily Law Reports Index*
... The Stationery Office *Daily List*	
... Journal source	
Checked to:	...... [latest dates]
European law	
General	
... *Encyclopedia of European Community Law*	
... *Common Market Reporter*	
... *Halsbury's Laws*	
... *Official Journal*	
... Other source	
....	
.... (name)	
Legislation	*Case Law*
... *Official Journal*	... *Common Market Law Reports*
... *Halsbury's Statutes*	... *European Court Reports*
... *Encyclopedia of European Community Law*	... Other law report
	.... [name]
... *Common Market Reporter*	
Supplement or service checked?	
.... [latest date]	
Updated by:	*To date*:
... *Official Journal*	....
... *EC Brief*	....
... *Halsbury's Laws of England noter-up*	....
... *Halsbury's Statutes noter-up*	....
... *Current Law*	....
... *European Current Law*	....
Forms and precedents; court practice	
Title	*Supplement or service date or date of publication* [*if not updated*]
... *Encyclopaedia of Forms and Precedents*	....
... *Atkin's Encyclopaedia of Court Forms*	....
... *Practical County Court Precedents*	....
... Other precedent book(s):	
Title	....
Title	....
Title	....
... Journal precedent (ref:)	....
... Database search (ref:)	
... White Book	....
... Green Book	....
... Stone	....
... Archbold	....
... *Blackstone's Criminal Practice*	....
... Other	....
....	....
.... (names)	....
Summary of research findings	
....	
	Advice (if appropriate)
....	

5.13 Practice exercise

(a) Two years ago Mrs Amery's mother, Mrs Ruth Ann Jacoby, executed a power of attorney in favour of Mrs Amery under the Enduring Powers of Attorney Act 1985. At the time Mrs Jacoby, who was separately advised in this matter, was in hospital following a stroke.

Mrs Jacoby is now fully recovered, but there has been a dispute within the family with respect to certain property, and our client, Mrs Amery wishes to disclaim the enduring power. I assume that this is possible, but could you ascertain whether there are any particular procedures which must be followed. Please draft any notices that are required under the Act.

Mrs Jacoby lives at Manor Farm, Upper Kentley, Barsetshire.

(b) As you know, this firm acts for Herb & Albert, architects. I have just received a telephone call from their managing partner. Since the summer of 1996, the firm has retained the services of a Mr Crick; he is a qualified architect who works freelance. He works for Herb & Albert three days a week in their offices on a self-employed basis. He has just written them a letter stating that, as a substantial part of his work involves him in the use of a word processor with a display screen, the firm must provide him with an eye-test. Please research and advise.

(c) Our client is Johnrose Ltd who lent £1,000 to H & B Hauliers Ltd to buy a ship 'The Lighten' secured by a registered mortgage. Our clients have now discovered that in April the ship was arrested by Barsetshire Harbour Authority for unpaid harbour dues and in August the ship was sold to Savage Investments Ltd for £4,000. Can we enforce the mortgage against the ship now that it has been sold? Please investigate the position and advise.

5.14 FURTHER READING

Clinch, P., *Using a Law Library* (London: Blackstone Press, 1992).

Dane, J., and Thomas, P. A., *How to Use a Law Library*, 3rd ed. (London: Sweet & Maxwell, 1995).

Price, M., Bitner, H., and Bysiewicz, S., *Effective Legal Research*, 4th ed. (Boston: Little, Brown, 1979).

Tunkel, V., *Legal Research* (London: Blackstone Press, 1992).

6

Practical problem-solving

6.1 Learning outcomes

This chapter deals with the skills needed for practical problem-solving. In this chapter we focus particularly on:

- understanding the importance of a problem-solving strategy for effective practice;
- using a variety of problem-solving techniques to:
  - (a) identify and manage factual information;
  - (b) define more precisely the nature of clients' problems;
  - (c) construct appropriate legal solutions to clients' problems;
  - (d) implement those solutions appropriate to your clients' goals;
- developing an effective risk-management strategy for your own work.

6.2 Why problem-solving?

As lawyers, we develop a range of professional knowledge, skills and values with a purpose and that purpose is to assist and advise those clients who come to us with their problems. Although we seldom think of it in these terms, effective problem-solving is thus at the heart of competent legal practice. This is reflected in the Law Society's LPC standards, shown below.

You should be able to:

(a) perform, with understanding, the skills and tasks required to complete transactions, in a manner which effectively achieves the client's and solicitor's objectives;

(b) demonstrate knowledge of and identify the rules of professional conduct and issues of client care in the relationship between solicitor and client;

(c) identify the client's objectives and different means of achieving those objectives;

(d) identify the steps and decisions that need to be taken to implement those objectives; and

(e) identify any difficulties that may arise in implementing those steps and procedures.

These standards are helpful in two ways. First, they stress the client-centred nature of legal problem-solving. As a solicitor your job is essentially to facilitate (so far as possible) your clients' objectives and desires, and all problem-solving is ultimately directed to that end. Secondly, by juxtaposing problem-solving and professional conduct, the standards also emphasise the key limit placed on that role: that you must not facilitate client objectives which would bring you into conflict with the law and with the accepted standards of professional conduct. We will make some brief references to professional conduct and client care in this chapter, but you are referred more generally to the relevant sections of the Law Society's Guide to Professional Conduct (on-line).

In this chapter, we treat problem-solving as a structured process, and have devised a 'process model' accordingly. This forms the basis for the discussion which follows.

6.2.1 A problem-solving model

The model adopted here breaks problem-solving into two phases involving six steps or stages, thus:

(a) Identifying the problem	
(b) Gathering the facts	*Definition*
(c) Defining the problem	*Phase*
(d) Developing options	
(e) Selecting the best solution	*Resolution*
(f) Implementing the solution	*Phase*

You may feel this is all pretty obvious and that you don't need something as structured as this to be an effective problem-solver. You could be right—to an extent! We all have many years of problem-solving experience and most of us have developed, albeit intuitively, at least some good practices out of that experience. But neither personal experience nor your LLB or CPE have necessarily prepared you for the problem-solving demands of modern practice. There are at least three reasons why you should consider adopting the kind of structured approach advocated in this chapter as shown in the box opposite.

6.2.2 The structure of this chapter

The remainder of this chapter is divided into seven main sections, the first six of these correspond to each of the steps in the problem-solving model presented above. In the final section we switch attention to the issue of your personal risk management. Professional practice inevitably involves an element of risk for the solicitor as well as the client. For the solicitor the key risks are of:

(a) failure to meet the reasonable expectations of the client;

(b) failure to meet the compliance and/or 'due diligence' requirements of the regulator; and

(c) failure to meet the standards of work expected by law (negligence, breach of fiduciary duty, etc.).

The profession has recently started to use terms like 'risk assessment' and 'risk management' to describe the need for each firm and each practitioner to monitor their own (best) practice. The aim of any risk assessment is to ensure that your standard of work meets both client and regulator expectations. The aim of this final section is to show you how this end may be achieved relatively simply and proactively.

Why we need a process model

(a) Legal problem-solving commonly involves finding solutions to complex problems, i.e., situations which involve 'problems within a problem'. Legal problem-solving therefore requires a cyclical rather than linear approach. By this we mean that problem-solving does not simply stop at step six (in our model), rather there is a problem-solving loop in which step six will often feed back into step one and re-start the cycle, as shown in the figure below. A more structured approach helps us to deal efficiently with that kind of complexity.

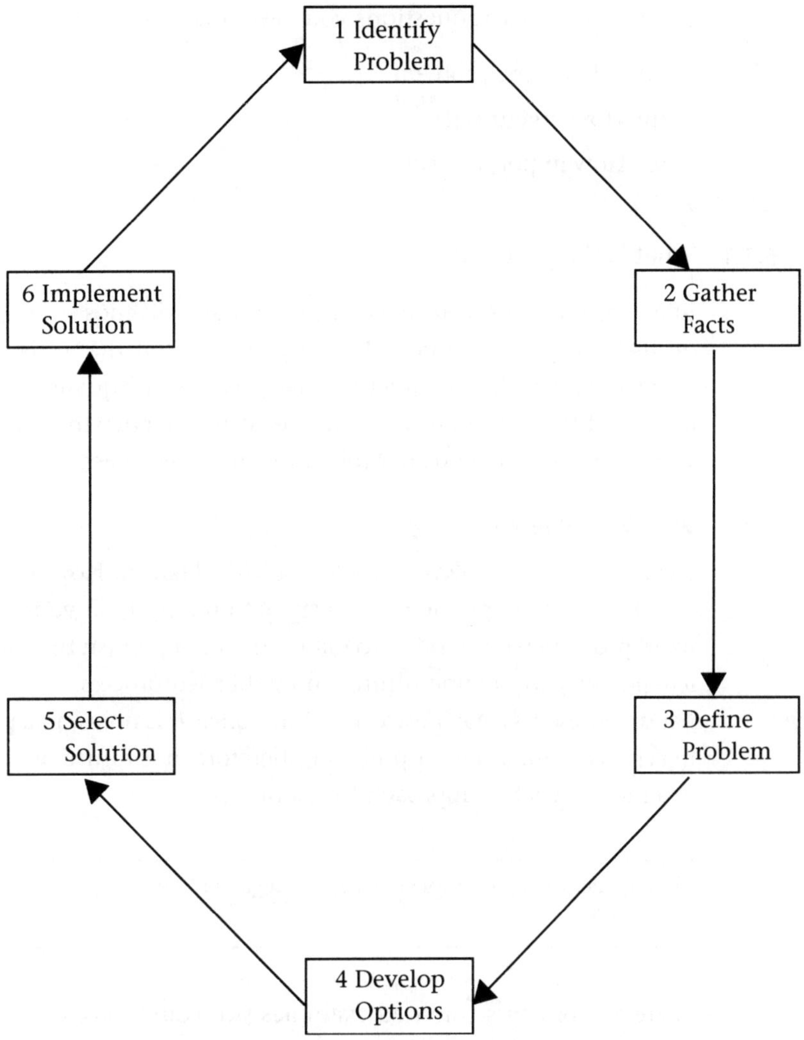

The problem-solving cycle

(b) Legal problem-solving involves problem-solving under pressure. You will inevitably be under time and cost constraints; you will certainly experience pressure from clients and even colleagues. It can be difficult to 'keep your head' in these stressful situations, in which case a simple, structured, model can act as an aid to clear thinking.

(c) You have a professional responsibility to be an efficient and an effective problem-solver. In most of our daily lives our decisions impact only on ourselves and those near to us. As a solicitor, you are holding yourself out to the public as a professional problem-solver. You will be answerable to your clients, your colleagues, and, ultimately, the courts and disciplinary bodies of the profession for the standards of work you maintain.

6.3 Step 1: identifying the problem

All legal problems involve some conflict between the way things are and the way your client wants them to be. This is equally true, for example, of whether you are trying to facilitate your client's house purchase, or their re-negotiation of a contract, or (hopefully) preventing them from going down for a lengthy prison sentence. But that does not mean that all problems are ultimately identical, and hence require identical strategies. This is why some degree of problem identification is a useful starting point. We suggest there are three basic questions you need to ask yourself:

(a) What is the problem?

(b) How urgent is it?

(c) How important is it?

6.3.1 What is the problem?

In all cases you need to make a preliminary 'diagnosis' of the problem, usually based on the information obtained from your client at the initial interview. Trite though it sounds, it is worth remembering therefore that the quality of your problem-solving will depend at the outset on the quality of your interviewing. In reviewing the information you have obtained, consider the following three questions.

6.3.1.1 What kind of problem is it?

Steven Kneeland (*Effective Problem-Solving* (Oxford: How To Books, 1999)) talks about 'Fix-It' and 'Do-It' problems as a way of thinking about your ultimate objectives. With 'a fix-it' problem we are trying to maintain or re-establish the status quo by making a problem go away, or by fine-tuning an established process. A 'do-it' problem, on the other hand, represents a gap between a situation as it is now and a preferred outcome we want to achieve. Unlike a fix-it problem, therefore, it is future-orientated, and may lend itself to a more proactive approach to resolution.

> Briefly identify an example of each of a legal fix-it and a do-it problem. In what key respects do they differ?

There are, of course, myriad examples you could have chosen. But the most basic distinction you will probably have observed is that litigation primarily involves 'fix-it' problems, while facilitative work (property negotiation, commercial contracts, etc.) is primarily 'do-it' work. A number of other key differences will have emerged, such as the degree of control you, or your client, may have over things like timescale, options and procedures, etc. We will look at these differences in more detail as we work through the chapter.

But noting that a problem is of a 'fix-it' or 'do-it' type doesn't by itself take us very far. We said at the outset that one of the features of practice is the extent to which it involves complex problems. The problem with complex problems is simply that they can have a range of dimensions which in turn could admit to multiple solutions. The important question is often, how do we know we are focusing on the right (part of the) problem?

Consider the following scenario:

> You have conducted an initial interview with a client, Ms X, who is a lone parent with a 6-year old child. She lives in a rented flat that she occupies as a secure tenant. Her landlord recently moved into the flat immediately below her after separating from his wife and leaving the matrimonial home. Your client thinks he may have significant financial problems. The property is damp and the fabric is in a poor state of repair and your client's relationship with her landlord has deteriorated dramatically since she started to complain about the worsening state of disrepair two or three months ago. Indeed, the landlord has apparently decided that your client is 'trouble' and wants her out. Consequently he has begun to 'harass' her by playing loud music late at night, shouting and banging on the ceiling, and she believes, dumping rubbish outside her front door. Your client has resisted leaving, partly because she doesn't want to give in to being 'bullied', and partly because she thinks it is going to be difficult to find another place to live that is sufficiently proximate to her workplace and her daughter's school, and that she can afford. At the same time she is also concerned that her and her daughter's health might begin to suffer.
>
> What would you advise?

A problem like this does not look factually complex, but it becomes more complex when you consider the range of possible solutions that could be available. These might include:

- *'legal solutions'* such as:
  - seeking an order restraining the landlord from continuing to harass your client;
  - enforcing express or possibly implied repairing covenants against the landlord; and
- *'non-legal solutions'* such as:
  - working with your client to convince the landlord to desist from harassing behaviour, and possibly working with the landlord and your client to find ways to address the disrepair problem;
  - assessing your client's eligibility for welfare benefits that might help her afford alternative housing;
  - introducing your client to a housing association or other provider of low-cost accommodation in the area.

Thus, one can focus on the problem primarily as a legal one of enforcement of rights, or a non-legal one, addressing your client's need for a healthy and non-threatening environment, which in turn could involve either fixing the current situation or enabling your client to move on to other accommodation. But however you define it, that will do much to determine the solution. So what's the best way of going about this? Here are some basic tips.

- **Look for the solution that is right for *this* client**—your client's needs and priorities should shape not just the solution you offer, but how you frame the whole problem. See section 6.5 below.
- **Be willing to think 'outside the (academic) box'**—a lot of academic training expects you to frame legal problems within established legal categories—contract, tort, etc. Practical legal problem-solving often needs you to cut across traditional boundaries to find solutions. See also section **6.6.2.6.**
- **Start by looking for the bit of the problem that is most easily solved or the solution that would make the biggest immediate difference to the client**—try not

to get bogged-down in the complexities of a whole problem. Break it down, and look for where you can make a difference for the client *now*.

- **Be practical**—a lot of legal problem-solving involves very little law; lawyering is often about finding non-legal solutions to apparently legal problems. In our example, this might mean focusing on the client's need for better accommodation and using your contacts in a local housing association to try to find her an alternative.

6.3.1.2 What are the causes, as against the symptoms, of the problem?

When a patient visits the doctor, the doctor will usually start by asking the patient to describe their symptoms. This is not because the doctor thinks it more important to treat the symptoms; rather, it is because by understanding the particular combination of symptoms they will be better able to diagnose the underlying cause, and treat that instead.

Legal practice is not so very different. To give a simple example, you are instructed by a client company to defend a claim of discrimination by one of its employees who has been refused promotion. That claim may well prove to be purely a symptom of a wider problem on which you ought ultimately to advise—perhaps there is direct discrimination by management, or evidence of an unintentionally but still indirectly discriminatory promotions procedures, or perhaps simply it is symptomatic of poor personnel management. This is also a simple example of how something that emerges as a 'fix-it' problem may ultimately be transformed into a 'do-it' problem. Like a doctor, you may start by treating the presenting symptoms, but ultimately you should be probing for the underlying cause. Steps two (fact gathering) and three (problem definition) of our model are therefore absolutely critical to the problem-solving process.

6.3.1.3 Have the client's objectives changed?

Client objectives often do not remain constant throughout a transaction. The nature of the problem may change with circumstances, or in the light of fresh information—including the advice that you give. It is important, both as a matter of effective problem-solving and good client care, that you keep your clients informed of developments and periodically re-visit what they want to achieve out of the transaction.

6.3.2 How urgent is it?

Every client would like to believe that their case is your top priority, but clearly this cannot be so. You need to assess for yourself the urgency of the work that comes across your desk. Your aim must be to try and avoid having to deal with everything as a matter of crisis management. Crisis management is a dangerous working practice because it encourages you to jump too quickly to action and increases the likelihood that you will make mistakes. This means that you must be prepared to stop and think, not just about the urgency of a task, but its importance in the overall scheme of your work: is it important enough to be 'now' urgent or 'tomorrow' urgent? We also discuss this as a more general issue of work management in **Chapter 10**.

6.3.3 How important is it?

Legal work involves a significant element of project management, of identifying the steps in the case or transaction and their relative priority. In assessing the importance of a particular task, it is often helpful to ask yourself the following three questions:

(a) Where does this task fit in relation to my client's objectives? Is this task central to what you are trying to achieve for your client; if not, what is its purpose? Do you need to do this at all? Even if it is important, you need to consider just how it fits in with your other objectives and tasks and seek to establish a sensible set of priorities. What this is will, of course, depend very much on the nature of each transaction.

(b) Is it the right time to address this issue? Legal work often has a linear structure built-in, so that there is a relatively logical progression from one task to the next—the stages in a residential conveyance, and the necessary pre-trial formalities of civil litigation provide obvious examples of this. You can use this structure to support your case management:
- Construct an action plan or checklist of key stages and critical dates and keep a copy in the front of the client's file (see section **6.7.1** below).
- Tick off steps as they are completed, so that you (and others) can quickly identify progress made on the file.
- Copy critical dates across to your diary or Outlook (and set reminders!).

  Your firm should also have a system of regular file audit and review in place as a means of monitoring progress and managing risk.

(c) Can I find a 'quick fix' that will contain the problem for now? Many transactions involve a mixture of short-term and longer-term objectives, and you need to identify and advise on both.

6.4 Step 2: gathering and managing the facts

Fact-finding and analysis are central to what lawyers do in practice. Unlike most law school problems, where the facts are presented to you pretty much 'on a plate', the quality of your casework in the real world depends far more fundamentally on your abilities. You need to ask yourself three things:

(a) What is the information I need to prove my case/complete this transaction?

(b) Where shall I find it?

(c) How can I make sure I use it to its best effect?

The first and second of these questions are essentially questions of fact investigation and analysis; the third depends on your skills of information management.

In our model fact investigation and analysis precede legal analysis. This is really a bit of an over-simplification because the whole relationship between fact investigation and legal analysis is symbiotic—each feeds off the other. We cannot pursue all the relevant facts without having some idea of the legal issue we are pursuing, but, equally, we cannot identify the legal issues without first obtaining some basic factual information. That is why we say fact investigation has logically to commence before any legal analysis.

What skills do you think contribute to effective fact investigation? Stop and make a note of all those you consider relevant before you continue reading.

We suggest the main relevant skills are:

(a) good interviewing skills;

(b) effective note-taking skills (as part of your interviewing);

(c) effective file and fact management skill; and

(d) a good understanding of the legal issues involved.

Taken together these will provide you not only with a good record of the relevant facts, and of the sources of those facts, but also a system for tracking the legal significance of that information. Interviewing skills are considered in **Chapter 2** of this guide; meanwhile, in the remainder of this section, we will look at a number of fact investigation and management techniques that are useful to legal practice.

6.4.1 Fact gathering and investigation

Let's start by thinking briefly about the main sources and categories of information you are likely to come across.

6.4.1.1 Sources of information

Although the potential range of information sources available to you in practice is both vast and dependent on the kind of work you specialise in, your initial information needs are likely to be met from among the following:

(a) your client—who is likely to be your first and often most important source of factual information;

(b) other participants in the matter—both lay participants (such as 'eye-' or 'ear'- witnesses; your client's employees, etc.) and those who are professionally involved (police officers, social workers, accountants, etc.);

(c) experts—particularly important in areas of litigation such as personal injury or professional negligence, but also in some kinds of non-contentious work—e.g., surveyors and valuers in the context of a range of property transactions, risk analysts in mergers and acquisitions work, and so on;

(d) other lawyers—be prepared to use both your colleagues and your opponents as an information resource: your colleagues because someone is almost certain to have been there before you (and may save you from re-inventing the wheel); your opponent because:

    (i) they can be obliged by discovery rules to disclose information that is of value to you in litigation, and

    (ii) particularly in non-contentious work, there may be a common interest in sharing certain information (don't always assume your opponent is your enemy!); and

(e) physical resources—you will almost certainly have to rely on many different kinds of documentary and real evidence: photographs, videos, a potential sea of paper documents, site visits, and so on. Effective management and use of this material is a critical part of any case management strategy.

6.4.1.2 Categories of information

It is always helpful to have some system which enables you to categorise the information you are seeking at a particular point in time. There are no hard and fast rules on how you go about this, though most solicitors in practice will use a system of checklists. Many practitioners' texts now provide specimen checklists that you can adopt or adapt for your own use.

Some areas of non-contentious work require their own very specialised checklists, reflecting the information needs of those fields. Otherwise, much of the information you will be obtaining is reasonably standard across most areas of contentious or non- contentious work. Although it is pretty obvious that you will not be asking exactly the same questions of a client who wishes to sell a house as you would ask of one who wants a will drafted, the broad kinds of information required are not that dissimilar. In contentious work, your initial information needs are probably more standardised. It is suggested that any checklist you devise is likely to incorporate elements of the seven information categories developed by Sherr (1986) 49 MLR 323 and listed in the table below.

Seven information categories

Personal information

Name, address, phone numbers, family ties, work, age, nationality, income and health may all be relevant—though not necessarily in all cases.

Other parties

Basic personal details; solicitor instructed (if any); connection with the client (if any).

Witnesses (if relevant)

Basic personal details (though often these will not be known by the client and must be followed up by the solicitor at a later date); witness to what and for whom? Connections with the client.

Events

Dates; times; place(s); persons involved; the cause and course of events; persons affected; property affected; precipitating incident to visiting the solicitor.

What the client wants

Identify the main problem; desired outcome; difficulties in achieving outcome; persons to be affected by outcome.

Previous advice and assistance

Anyone else consulted? Details of consultant; the advice given; action taken by consultant and by client; effects of any action taken. (In Sherr's research this and the category following were the areas least often addressed by trainee solicitors when interviewing.)

Existing legal proceedings

Nature of the proceedings; parties; stage of process; past/future hearing dates.

You will be given sample checklists and aide-memoirs and/or encouraged to develop your own while on the LPC.

6.4.2 Fact analysis

As we have already said, facts are not something that are just presented to you, you have to develop a deep understanding of the information in your possession, and be able to identify important gaps and contradictions in that information. Good fact analysis therefore requires:

(a) understanding the 'context' in which the problem arises;

(b) knowing the 'standpoint' from which you are approaching a problem; and

(c) the ability to 'work' the information at your disposal (the critical skill of what we call 'fact appreciation').

6.4.2.1 Understanding your context and standpoint

Legal problems don't arise in a vacuum, they arise out of real human activities. Problem-solving thus requires both an understanding, as we have said, of the legal issues within which you will seek to frame the problem, but also of the (non-legal) context. If you do medical negligence work, for example, you need some appreciation of the administrative and medical procedures which framed or created the circumstances in which the complaint arose. If you do mergers and acquisitions work you need to know the nature of the particular market; you need to be able to understand the input of other professionals into the deal (so that you can make sense of the outcomes of the due diligence process, for example), and so on. Obviously your understanding of a particular context will increase with experience, but it does mean that, in the early days of practice you must get into the habit of thinking about the business or other context within which your work will arise—for example, by reading the relevant financial or trade press and by using your colleagues, clients and others as an information resource.

We suggest thinking about standpoint is important because it encourages you to consider the facts that you need and how and why you want to use them. Twining and Miers (*How to do Things with Rules* 4th ed. (London: Butterworths, 1999), p. 77), suggest that 'standpoint' boils down to asking yourself three questions:

(a) Who am I?

(b) At what stage in what process am I?

(c) What am I trying to do?

Try asking these questions next time you are analysing a problem, and see if it helps you focus any differently on your information needs.

6.4.2.2 Fact appreciation

Information gathering is largely a matter of asking the right questions in the right way, a process which itself depends largely on your skills of fact appreciation. If you don't understand what questions you ought to be asking, and don't analyse critically the information you have, the quality of your information gathering is likely to remain poor. Appreciation techniques are for the most part pretty obvious, and straightforward. They include:

The six basic questions

When you are presented with a piece of information, whether you are interviewing a client, or a witness, or just mentally interrogating a documentary source of information, it is always useful to keep in mind the six key words you can use in framing questions about that information:

(a) Who?

(b) What?

(c) When?

(d) Where?

(e) Why?

(f) How?

By using these keywords repetitively and in various combinations, you can generate some powerful techniques for interrogating the information (but hopefully not the person in possession of it!) at your disposal. Again, we know this sounds terribly simple and logical, and you may be thinking 'I don't need this' but experience shows that an awful lot of mistakes are made simply by a failure to ask the right questions or to read documents carefully and critically. This is well illustrated by the following anecdote.

> I was acting for a truck driver in a personal injury case. It was a garden-variety whiplash. The client was rear-ended... He brought in a big pile of documents—pay slips, claim forms, medical reports, physiotherapy reports, a messy diary and so forth. I read and organised them, noticing that he had made a previous whiplash claim and received $7,600. We discussed this and he explained that it had occurred four years ago and he had completely recovered. I asked him if he had made any other claims and he said 'no'. I got the file going, started legal action and notified the defendant's insurer.
>
> [The defendant subsequently made a payment into court of $12,000 which, contrary to advice, the claimant rejected.]
>
> A year later we went to trial. During my client's cross-examination, he admitted the previous whiplash claim and the $7,600 settlement. Then counsel dropped a bombshell. 'Did you make yet another similiar claim five months after you made the first one? Counsel then produced a claim form with my client's signature on it. I looked through my document bundle to find the form and sure enough, a copy of it was there. I had had it all along. The insurer's solicitor had given me a copy in discovery. But it was so similar in form and content to the first form, and so close in time, I had assumed—wrongly, of course—that it was simply another form filled out for the same claim.
>
> But then I looked at it more closely and I noticed it had a different claim number! It was another claim for another accident and, to the court, it must have made my client look like a con man. My heart sank and so did my client's credibility. His case collapsed. I considered myself lucky when the court was decent enough to award him $2,800 even though it was not enough to pay the defendant's costs.
>
> From: Nathanson, S., *What Lawyers Do: A Problem Solving Approach to Legal Practice* (London: Sweet & Maxwell, 1997), pp. 87–88.

Had the solicitor here simply asked why there was what appeared to be an unnecessary duplication of a claim among the documents disclosed, then he might have gone back and examined the 'duplicate' more closely.

'So what?' and 'repetitive why'

These are two particular examples of the kind of technique we have just been talking about. By starting with a known statement of fact and asking 'so what', you can work out the implications to be drawn from that fact. By repeatedly asking the 'so what' question you can quickly and easily ascertain a range of inferences or implications which flow from any given fact. This can be useful in ascertaining evidential or other information needs. To give a simple example:

Fact: it rained heavily last night:

(a) So what? The ground will be wet.
(b) So what? It will turn to mud easily.
(c) So what? Any vehicle driving over soft ground is likely to leave tyre tracks.
(d) So what? We should check whether tyre tracks were found and, if so, see how they compare with the tyres on our client's vehicle.

A variant of this technique, which may be particularly useful for dealing with 'fix-it' problems, is to ask the 'repetitive why'. Unlike 'so what' which is forward looking, in the sense that it enables us to work out the implications of a fact, 'repetitive why' helps us track back and uncover the root causes of a problem. It works by identifying a manifest symptom (e.g., 'the other side refused to pay for the goods my client delivered') and working back by repeatedly asking 'Why did that happen' or 'what caused that'.

Charting unknowns

Identifying gaps in a fact pattern can be an important activity when developing our understanding of a problem. A very simple technique is to ask: 'What don't I know about this problem'. This can be useful, particularly with difficult fact patterns, where mental blocks sometimes occur because we get fixed on what we know about the problem, rather than what we don't. Having identified the gaps, make a list of them, and use this to inform a further phase of fact investigation and analysis.

6.4.3 Fact management

As lawyers we are frequently involved in tasks that are factually complex. We therefore need to develop effective techniques for managing large amounts of information. This may be especially critical in litigation, where information needs to be organised in such a way as to make it readily useable and understandable in court and under pressure. Again there are various tools that can help you in this process:

6.4.3.1 The chronology

A chronology lays out facts and events along a timeline. This can be helpful for a number of reasons:

(a) most obviously, it helps you understand clearly the sequence in which events happened;

(b) it can assist in identifying gaps in that sequences;

(c) it may provide you with a structure for presenting the case in court; and

(d) outside of litigation, it may help you meet project and time management targets.

The structure is very simple, requiring at a minimum just two columns, the first noting the date and time of an event, the second stating briefly the event itself. This might also be incorporated into the next technique we discuss—the trial notebook—particularly if you intend to adopt a chronological as opposed to topic-based scheme for presenting your case.

6.4.3.2 The trial notebook

A trial notebook is a technique for providing total management of information in a case. The notebook itself can be pretty well whatever you want it to be—a single notebook, a ring binder, or a set of notebooks or files—which contains your notes and supporting documents on a case and is organised into a methodical set of materials which can be used in court.

Within the notebook you should have a copy of your working theory of the case and, whenever possible, a separate analysis of your opponent's case. The main mechanism for gathering and managing the evidence is a proof checklist. This will contain a record of evidence divided into three columns, thus:

Ingredients to be proved	*Supporting evidence*	*Source of evidence*

The organising principle is topical: you focus first on what you need to prove to succeed, for example, the need to prove the terms of a contract, compliance with those terms by the claimant, and a breach by the defendant which caused the claimant to suffer damage. The second column will then list the specific supporting evidence, and the final column will identify its source, together with a note of its page number in any agreed bundle of documents.

This is the simplest record to compile, but it does tend to overlook the conceptual differences, and the need to construct probative links between the ingredients (a matter of law), the facts and the evidence. Particularly in the evidentially more complex case, these distinctions may need to be followed through precisely—where, for example, one ingredient can only be proved by an accumulation of evidence supporting several facts in issue. It would of course be possible to insert an additional column between ingredients and supporting evidence to list specific issuance or collateral facts, but this would make the checklist significantly more cumbersome to construct.

The rest of the 'notebook' will contain the documentary evidence in the case and any research notes and skeleton speech you have prepared. All this material should be paginated or colour coded to distinguish the different sections.

6.4.3.3 Statement analysis

While a chronology or proof checklist provides you with a route map through the case, it will not ensure you have sufficient detail to prove all the facts required to put your case. Statement analysis provides a thorough method of checking that all necessary items of evidence have been proved in court.

Statement analysis involves just three steps (see Sherr, A., *Legal Practice Handbook: Advocacy* (London: Blackstone Press, 1993), pp. 35–6):

(a) Reduce each page of your witness statements or proofs of evidence on a photocopier so that it leaves a substantial margin to one side of or around the statement (though make sure the text of the statement remains legible).

(b) Highlight or underline the key words or sentences in the statement identifying the facts you need to prove in chief.

(c) In the blank margin you have just created, number each separate point you need to prove.

The result should look something like the Statement Analysis shown below. This kind of analysis can be readily incorporated into the documentary section of your trial binder, so that, when conducting an examination-in-chief, you can then tick off each item as it is stated by the witness.

Statement analysis

Proof of Evidence

GARRY JONES will say:

I live at 44b Cheap St, Bow London with my mother and younger brother. I am aged 17 and still at college. I understand that I have been charged with aggravated vehicle taking along with two others. I intend to plead not guilty as I was an innocent passenger and I cannot believe I have been had up for this.

Facts:

On July 2nd, 2000 after my tea I decided to visit a friend, Justin Rafferty, who lives down the other end of the Bow Road from my mum's. (1) ◂ I left the flat at about 8.15 p.m. About 10 minutes later, as I was walking down the Bow Road, this red Ford Escort stopped. (2) ◂ I did not know the driver, but in the passenger seat was a friend from college, Danny Lu, who asked where I was going. I said I was off to see a mate who lived at the other end of the Bow Road. (3) ◂ Danny said to jump in and they would give me a lift, so I got into the back of the car. (4) ◂ Danny then introduced me to the driver, who, he said, was a mate of his called Lee. I now know this to be Lee Evans. Danny and Lee were sharing a spliff, which they proceeded also to share with me. Danny said they were planning to head up the West End for some fun and did I want to join them. I figured Justin wouldn't mind, so I agreed. We can't have been in the car more than another ten minutes when we heard a siren and saw a police car coming up behind. Lee just took off up the road, through a

6.5 Step 3: defining the problem

Once you have commenced fact investigation you can also begin to refine your definition of the problem. Problem definition is about understanding the problem, it involves:

(a) understanding the facts of the matter;
(b) understanding your client's goals;
(c) understanding the goals of other parties to the case or transaction; and
(d) translating this information into a 'legal' problem frame.

6.5.1 The elements of problem definition

There are two phases to problem definition. The first is to review the problem itself—to look at how things are now—which requires us to be satisfied that we have a grasp of both the symptoms and causes of the problem (see **6.3.1**, above). The second phase is, on the basis of that analysis, to begin to predict a range of possible solutions—the way we want things to be. Here there are two key criteria that any preferred solution must satisfy. These are the outcomes which we want our solution to achieve and the constraints which limit the range of practicable solutions. Problem definition thus involves the accurate identification and mapping of these four characteristics—symptoms, causes, outcomes and constraints (SCOC)—see below.

SCOC

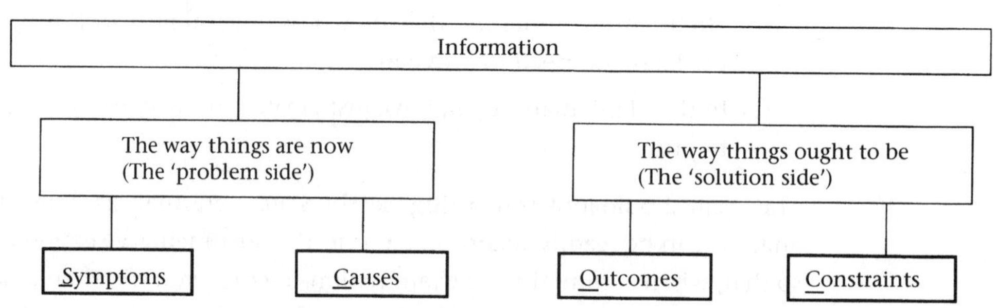

(From Kneeland (1999), p. 53)

The relative importance of these phases is largely determined by the kind of problem we are dealing with.

6.5.1.1 'Fix-it' and 'do-it' problems

If our problem is a 'fix-it'-type problem, Kneeland suggests most of the time we spend on fact analysis and problem definition is likely to be spent on identifying symptoms and causes; if it is a 'do-it' problem, the focus will need to be more on outcomes and constraints. As we have already looked at the problem side of the equation, we shall concentrate here on the solution side.

6.5.1.2 Outcomes and constraints

In principle the outcomes are determined primarily by your client. You should remember that although Practice Rule 1 of the Solicitors' Practice Rules 1990 requires you to act in the best interests of your client, this does not empower you to override express instructions from the client simply because you do not believe them to be in their best interest. Your duty under a retainer is certainly to advise your client as to any negative consequences or risks inherent in their instructions, but ultimately you are obliged to carry out those instructions diligently, unless you have grounds on which to withdraw from the retainer. In practice, of course, this means that the outcomes of a transaction are essentially negotiated between solicitor and client. Often (though not inevitably) they are a compromise between what the client really wants and what the lawyer thinks realistically can be achieved. What can be achieved will depend on a range of constraining factors.

> Before reading on, make a quick note of the kinds of factors which you think will commonly act as constraints on the range of potential solutions to a problem.

We suggest key constraints are likely to include:

(a) the facts available—particularly in contentious matters, where what you can achieve will depend, to some extent, on what you and the other side think can be proved;

(b) the range of acceptable solutions: certain solutions may not be acceptable to your client; some may not be achievable within the law. Again, this is not least because you are ethically constrained only to advance lawful solutions to clients' problems;

(c) the goals of other parties, since these will also help define the outcomes that are achievable; and

(d) material constraints such as time, money, human resources.

6.5.2 Constructing the legal problem frame

Legal issue identification and analysis is central to the task of problem definition, because it facilitates the translation of the 'lay' problem into a 'legal' solution.

In working through a legal problem analysis you need to remember that, as we have said, the relationship between fact and law is symbiotic, so, while your legal analysis is triggered by the factual information you have established, it also underpins the process of fact analysis. The need to establish a clear linkage between law and fact is especially great in contentious work because you are obliged to prove your case on the basis of evidence which shows your claim (or defence) to be made out in all material particulars. This means that, in identifying an appropriate legal framework, you need to identify three things:

(a) a legal right or obligation which is prima facie supported by the facts of the case (e.g., rights accruing under a contract, will, etc., a criminal offence);

(b) a cause of action accruing from that right (e.g., an action for damages for breach of contract, or in a criminal matter, we would look to the particulars of the offence); and

(c) the ingredients of the cause of action, etc., by which we mean the elements of the cause or offence which must be proved by evidence—e.g., in a theft case the prosecution would need to prove (unless any element was formally admitted) all the elements of theft, i.e., a dishonest appropriation of property belonging to another with the requisite intent.

This process of analysis can be represented graphically, using a system developed by the College of Law in Sydney, Australia. This technique is variously referred to as 'five level analysis' or sometimes 'charting'. The strength of the charting method is that it enables you to represent your legal and factual analysis conjunctively. Although there are some claims that charting can be adapted effectively for non-contentious work, it is our view that it is primarily a case analysis tool for use in contentious matters, and that is how we will represent it here.

6.5.3 Charting litigation

Charting works by identifying the five different levels of information you will encounter in a legal problem. It does this by starting the analysis with the relevant law at levels 1 to 3 and then, in levels 4 and 5, mapping the issues of fact and actual items of evidence against the legal requirements, thus (taking a simple contractual dispute as an example):

Level 1	Source of client's/Crown's right	*e.g., Sale of Goods Act 1979 (as amended)*
Level 2	Cause of action/charge	*e.g., action for damages for breach of implied terms*
Level 3	Ingredients of action/charge	*i.e., what elements the claimant/prosecution must prove to make out their case—e.g., that there was a contract, that s. 14, SOGA applied, etc.*
Level 4	Propositions of fact in support of each ingredient at level 3	*e.g., that a contract existed, that goods were delivered, that the goods were not of satisfactory quality, that loss arose as a consequence, etc.*
Level 5	Items of evidence which prove	*e.g., the written contract, parole evidence of facts alleged at level 4 checks made at time of delivery, etc.*

This relationship can be best represented graphically, as this enables us to track the relationship between each element of law, fact, and evidence, as follows:

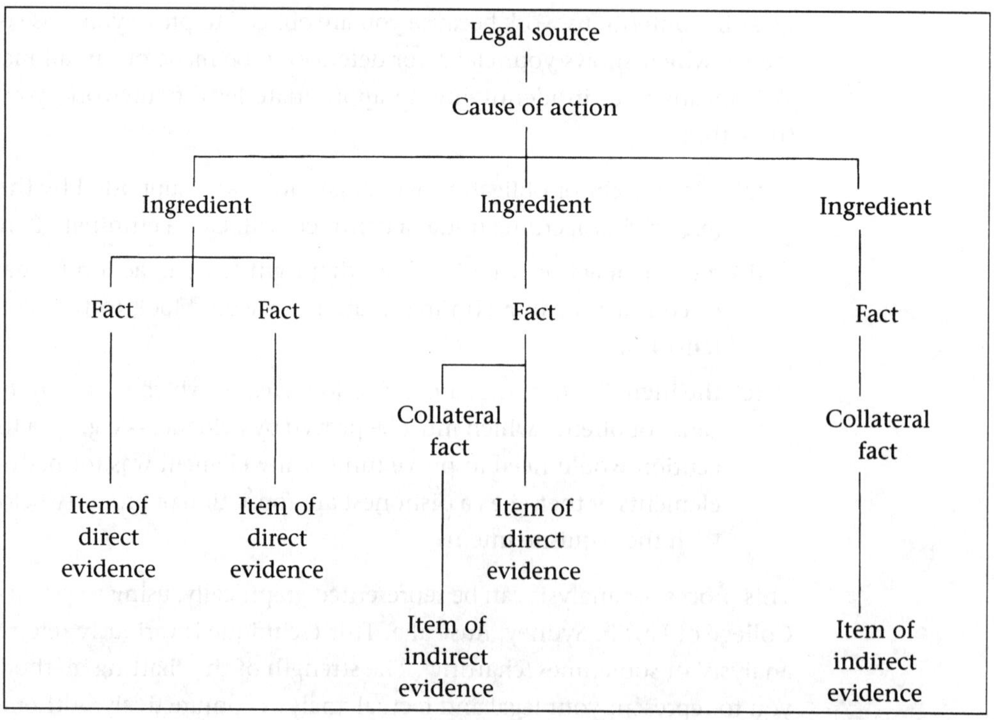

It may also be desirable to attempt a chart analysis of your opponent's case. This enables you to note facts which, for example, are not admitted or are traversed by the other side, and thereby helps to highlight the points in dispute, and the nature of that dispute. For the sake of clarity, the analysis of an opponent's evidence is best done on a separate chart rather than superimposed on the chart of your own evidence, though the two could be presented in parallel on the same sheet of paper—provided you have a sufficiently large sheet, or can draw very small!

6.5.4 Assessing the legal issues

Techniques like charting are useful as a mechanism for identifying the relationship between fact and law, and as part of the process of legal issue identification. But we also ned to move beyond legal issue identification. Part of the practitioner's skill as a problem-solver is not just to identify but to assess the legal issues. That is, we take on the task of assessing the kinds and merits of legal arguments that might be raised on either side of an issue to enable us to decide what direction our problem-solving is going to take. 'Do-it' and 'fix-it' problems tend generally to require different kinds of assessment.

6.5.4.1 'Fix-it' problems

'Fix-it' problems (e.g., when litigating or negotiating a claim) involve what Nathanson calls 'prediction assessments'—assessing how a court would deal with the legal issues raised in the context of those particular facts. This is a skill with which we are all reasonably familiar (albeit in a fairly abstract form) from law school. The key difference, of course, is that in practice, unlike law school, the prediction assessment doesn't finally determine your course of action. Although a prediction assessment is a necessary and useful tool, you will have to take into account a range of strictly 'non-legal' issues in advising on a course of action, and it will often be the case that these non-legal issues are far more determinative of the outcome.

6.5.4.2 'Do-it' problems

These, on the other hand, often require a different kind of assessment—what Nathanson calls 'credible position assessment'. Don't let the jargon put you off. The distinction is a useful one. What you are asking is this: on a specific legal issue, can my opponent maintain a credible position arguing for the other side? Obviously if they cannot, this means you have identified a potentially strong bargaining position, if they can, then asking the question has helped you to identify the need to limit or block the potential for future conflict. We can illustrate this by using Nathanson's own example (at p. 43).

Assume you are acting for the prospective purchaser (Ms P) in a conveyancing transaction. Mr S is the sole legal owner of the house Ms P wishes to buy, and no other beneficial interests are disclosed by a search of the Registry. Ms P however tells you that, when viewing the property, she met a woman (Ms B) who also appeared to be in occupation. Consequently, you are concerned not only that Ms B has an unregistered beneficial interest, but that her interest could, on these facts, survive your client's purchase of the legal estate. If you accept that this is an unlikely but nevertheless legally credible position for Ms B to take, then, to protect your client from the risk of future conflict, you could advise Ms P to seek either a release from Ms B or at least confirmation that she has no such interest. This example rather neatly leads us into the next phase.

6.6 Steps 4 & 5: developing and selecting solutions

In practice the best option for a client can fall anywhere on a continuum from the blindingly obvious to the extremely elusive. At one extreme there may be little need to evaluate options. There are many occasions when it is unnecessary or impractical even to look for a range of options; rather the problem-solving process is dominated by the developing and refining of a single solution. The tendency for law firms to standardise the more routine work through the use of know-how systems and increasingly sophisticated case management tools, has taken a lot of the effort, guesswork and, indeed, some of the expertise out of these stages of problem-solving. At the other extreme, however, selecting and evaluating options is a highly complex and technical task. In non-routine cases these can be the most challenging stages in practical problem-solving, and the most heavily dependent on experience and intuition. It is not surprising therefore that one of the realities of life as a trainee, and often as a newly qualified solicitor too, is that you will spend a lot of time as an observer rather than active participant in this stage of the process. For this reason we focus here more on the interpersonal than the strategic aspects of discussing options and selecting solutions. We have organised the discussion around three general principles that should help inform this part of the process.

6.6.1 Principle 1: decision-making should be participatory

As a grounding principle it is generally accepted in both theory and practice that lawyer-client decision-making is a shared process. Now, that still rather begs the question as to just how involved clients should be in decision-making. There are three helpful rules of thumb:

6.6.1.1 Leave 'big decisions' to the client

Current best practice suggests decisions which affect the outcomes of a transaction should be discussed with the client; process decisions as to how the matter is handled are more likely to be strategic issues which are within the solicitor's own remit (see Nathanson, *What Lawyers Do*, 1997, p. 46; Sherr, *Advocacy*, 1999, p. 114). James Freund (*Lawyering: A Realistic Approach to Legal Practice* (1977)) has come up with a catchy way of remembering this as a distinction between 'go/no-go' and 'way to go' decisions. Note however that as a statement of general principle this goes beyond, but is not inconsistent with, the client care and other general obligations to consult clients which are contained in the *Guide to Professional Conduct*.

6.6.1.2 Professional conduct rules may require it

Exceptionally client consent (rather than just consultation) may be expressly required in respect of certain decisions. Most examples relate to 'process' decisions that would create potential conflicts of interest between clients (e.g., where a solicitor proposes acting for both the seller and buyer of residential property, in the limited circumstances where this is permitted, or when appearing as an advocate for co-defendants).

6.6.1.3 If in doubt, ask the client

In many cases there are no rules, or their scope is uncertain. Also, different clients will have different ideas about how involved they wish to be in decision-making. It may therefore be advisable to obtain instructions from the client as to the kind of consultation she or he expects, and to confirm this in writing.

6.6.2 Principle 2: decision-making will benefit from structured evaluation

Assuming your problem does not come with a ready made solution, you will need to find some mechanisms for developing and evaluating options for presentation to the client. Below is a list of techniques that could be used to identify and prioritise options. Not all of them are commonly used in practice, though they all have that potential. We have ordered them from the least to the most structured:

(a) informal discussion;

(b) brainstorming;

(c) weighing against objectives;

(d) weighing against consequences;

(e) evaluating against criteria; and

(f) excluding bias and fallacy.

We can now consider each of these in turn.

6.6.2.1 Informal discussion

Our conventional image of solicitors portrays them as working on their own case load in virtual isolation from colleagues. This is, and always has been, something of an over-simplification, but it is also true that the increasing complexity of practice (especially corporate work), and the consequent need for specialisation has greatly increased the incidence of team-working and collective decision-making in law firms. Consequently planning and evaluation meetings have become a more commonplace feature of legal work. These may be more or less structured, depending on the issues and preferences of team members/leaders.

Even where decision-making is down to a single fee-earner, informal discussion with colleagues can be an invaluable way of talking through ideas and solutions. For a new practitioner, in particular, the opportunity to draw on the experience of more senior colleagues is a critical part of developing casework skills and expertise.

6.6.2.2 Brainstorming

Brainstorming is not really a complete method for developing solutions, but it can be a powerful tool for getting the process going, particularly where a problem requires a creative rather than 'off the peg' solution.

Brainstorming works best as a group activity. The aim is to maximise the number of options/ideas we can produce about a problem and to encourage creative thinking through the momentum of the group activity. The basic technique is very simple: the group should focus on the issue or problem and individuals should be encouraged simply to offer up ideas for its resolution, without stopping to evaluate those ideas. While there is the risk that some of the ideas you generate will prove wholly impractical in themselves, they may act as a trigger for other, more workable solutions. A good brainstorming session should produce some useful options, which the group can then move on to evaluate more objectively.

6.6.2.3 Weighing against objectives

This is somewhat self-evident, but it is important to check any options against the client's objectives. Other things being equal, the option which comes closest to meeting those objectives should be preferred. However, it will often be the case that other things are not equal, and your decision-making therefore demands some trade-off between the optimum and most practicable solution. In this case, some further evaluation against consequence or criteria may be necessary.

6.6.2.4 Weighing against consequences

By identifying the possible consequences of a range of solutions we can evaluate each solution in terms of the risks/rewards and costs/benefits accruing to the client. These can be mapped out on a consequences work sheet (Pokras, S., '*Successful Problem-Solving and Decision-Making*, 2nd ed. (London: Kogan Page, 1990, p. 95), thus:

Consequences work sheet

Solution	Potential costs	Potential risks	Possible benefits	Possible rewards	Conclusions

In the conclusion column you should indicate whether the benefits and rewards justify or outweigh the potential risks and costs.

Risks and rewards

In this context, we are concerned with the risks to which your client is exposed by a particular problem and a proposed solution. These risks are somewhat different from those considered at **6.8** below ('personal risk management'). In assessing your clients' risks it is useful to ask yourself the following questions:

(a) What can go wrong?

(b) How likely is it that it will go wrong?

(c) How serious would the consequences be for the client?

(d) Can we limit the risk within the framework of this solution?

(e) What are the consequences of limiting that risk?

The connection between risk and reward is often linear. High risk strategies frequently carry higher rewards, but the risk of failure tends also to be high. For example, if you do defendant personal injury work, and you approach the negotiation of a specific claim in a highly competitive manner, you may achieve a higher settlement than the average cooperative negotiator, but the risk of negotiations breaking down is also greater than if you (and your opponent) take a more cooperative approach. This kind of equation needs to be identified and, perhaps, discussed with the client.

Costs/Benefits

Constructing a proper, quantitative, cost/benefit analysis can be quite a sophisticated task, but it is within anyone's capabilities to construct a relatively qualitative comparison of costs and benefits of a course of action.

Assume, for example, you are advising on a commercial contract dispute. Your options may be to settle by negotiation now; to mediate the claim within the next six weeks, to arbitrate within the next three to four months, or to litigate at some point thereafter.

We could analyse this by looking primarily at the costs of sustaining the dispute over the necessary timescale likely to be alloted to each option. Following Levine (*Getting to Resolution: Turning Conflict into Collaboration*(San Francisco: Berrett-Koehler Publ., 1998), p. 16), we might identify the following categories of dispute cost:

(a) Direct costs—fees for lawyers, courts, other professionals (mediators, arbitrators, experts, if required) etc.;

(b) Productivity costs—value of time lost (e.g., by managers, in-house lawyers, etc.) to servicing the dispute which could be used more productively;

(c) Continuity costs—the damage to ongoing relationships in the organisation/with other businesses/with the community, e.g., loss of trust; damage to communications networks, damage to continuing contractual relations, etc. (this category may have to be treated as qualitative as such costs may be difficult to quantify); and

(d) Emotional costs—the need to account for the stress and emotional pain of differing conflict situations (again, an essentially qualitative assessment).

For a number of our options, particularly negotiation and mediation, the key benefit is that many of these costs are likely to be lower than they are for the more formal alternatives of arbitration or trial. But at the same time other factors, like the merits (or otherwise) of having a public/formal as opposed to private/informal resolution of the dispute, also need to be factored into the cost/benefit analysis.

6.6.2.5 Evaluating against criteria

A rather more sophisticated method of weighing solutions is really to combine the previous two models so that we weigh solutions against a set of criteria that are agreed with, or at least known to be important to the client. Each of these criteria must be an accepted benchmark or standard which shows that the solution is likely to be acceptable or effective. To give a simplified example, in mergers and acquisitions work, you might agree that the criteria against which you assess the decision to proceed with one takeover rather than another are: the likely impact on share price, impact on market share, anticipated economies of scale, cost of regulatory compliance, effect on management and staff morale in each organisation, and so on. Each of these criteria can then be rated against a uniform scale—this could be numerical in some cases—cash values, say—or it could be more qualitative, for example, by simply rating options as having positive (+) negative (−) or uncertain (?) outcomes on those criteria. This kind of analysis can also be developed in a work sheet format—see, e.g., Pokras (1990) p. 101.

6.6.2.6 Excluding bias and fallacy

In weighing solutions you should always guard against the risk of adopting a solution for bad reasons. You should evaluate your own decision-making and check for biases and fallacies.

Bias is a problem that can creep up on any of us. One of these is quite fundamental, consider: what kind of lawyer are you? The psychologist Robert Redmount (quoted in Hegland, *Trial and Practice Skills*, West Publishing, 2002) argues there are basically three types of lawyers:

- those who seek to improve their clients' *economic* position;
- those who aim to vindicate their clients' *rights*;
- those who seek to reduce their client's *conflict* positions.

The point is that none of these are objectively right or wrong, and all of them can serve the needs of different clients, but one of them, as a standard approach may not serve the needs of all clients equally well. As Hegland puts it, 'predispositions are like glasses, they bend the incoming data'. Think about what your own biases are. Discuss cases with colleagues, to see how their perspective and approach differs from your own.

Fallacies can be logical or psychological assumptions and biases that trick us into bad reasoning and decision-making. There are many of these, but it would get too complicated to explore them all; some key examples that can impact on practical legal problem-solving are:

- *The fallacy of availability*—we tend to give priority to what we already know and understand. It can be nothing more than a tendency to stick by tried and tested solutions because they work and we understand them, and the value of that should not be disregarded. But, at the same time, this approach means we don't confront what we don't know, and do not necessarily consider the possibility that there may be a better alternative out there somewhere. People who defended the theory that the earth was flat against the growing scientific evidence to the contrary operated largely on the basis of this fallacy.
- *The anchoring fallacy*—anchoring consists of a failure to adjust one's 'world view' to accommodate new evidence, and the tendency to try and explain away anomalies rather than amend one's theory in the light of them. We tend to hold onto established ideas and cherished theories, often well beyond their 'sell-by' date. In law the failure to adapt your theory of the case to accommodate new evidence would be an example of this fallacy at work.
- *The recency fallacy*—the obverse of anchoring is that sometimes we place too much emphasis on recent events simply because they are recent, and so more prominent in our memories. The fact that your favourite football team has lost six matches in a row may lead you to conclude that you are having a really bad season, overlooking the fact that you also won ten of the previous eleven games. Again, to give an example in a legal context, the recency fallacy could lead you to attach too much weight to the most recent evidence you have obtained—particularly if it is favourable to your case—where recency and anchoring effects might actually work in tandem.

The problem of bias and fallacy in legal reasoning and decision-making is discussed in greater depth in David Crump's *How to Reason About the Law* (LexisNexis, 2001), Ch. 2, from which some of the above examples are drawn.

6.6.3 Principle 3: consider how you present solutions to clients

If problem-solving is to be a collaborative process between lawyer and client, you will need to consider how you present your options to your client. In short, you must give your clients a structure they can use to assess the options and come to a decision. Usually the simplest technique is to work through a logical process of elimination, taking out those options which are not workable, e.g., because they are too risky, or too expensive, or do not sufficiently meet the client's primary objectives. Do also remember that part of your role, even at this stage, may be to help the client face the fact that certain of his or her objectives are unachievable. Presenting clients with a range of options can sometimes be a useful way of forcing them to confront this reality and to re-prioritise.

You should also put some thought into the way you intend to present your solution(s) to the client. Should you send the client a letter of advice before any meeting, so that the client has had time to think about the issues you have raised? Have you considered who your audience will be? Is it a single client, a number of your client's representatives, or are you providing the legal advice as part of a multidisciplinary team of professional advisers? Each of these contexts may require different levels of formality, different styles of presentation, and so on. Does the advice you are giving lend itself better to one form of presentation than another? Complex information is often best presented visually, so think about using flowcharts, work sheets (as in **6.6.2.4** above) and similar presentational techniques to aid comprehension.

6.7 Step 6: implementing the solution

Agreeing on a solution does not of itself resolve a problem. We have to translate that decision into a plan of action:

(a) What, generally, are the main issues that will need to be resolved in translating a solution into action?

(b) Do you think the solicitor's role in implementation might differ as between 'fix-it' and 'do-it' problems?

Your answer to the first question should have picked up the following:

(a) What specific tasks are involved?

(b) Who is responsible for managing and who for undertaking each task?

(c) What is the schedule for:
   (i) the implementation phase as a whole; and
   (ii) each stage of implementation?

(d) What resources are needed?

As regards the second question, this is less easy. Your role in implementing solutions will vary from task to task and client to client. Many (though by no means all) 'do-it' problems will require little lawyer involvement at this stage. Rather, your role will have been primarily to identify the best legal solution in the circumstances and brief your client accordingly. For example, if you have been assisting your client to develop appropriate criteria and procedures for a redundancy, the actual implementation of those criteria will be your client's responsibility not yours. By contrast, the responsibility for implementing most 'fix-it' solutions will remain with you, the solicitor. It is important therefore that you try to ensure implementation proceeds in a planned and organised fashion. This is the function of 'action planning'.

6.7.1 Action planning

Action planning techniques are widely used in business contexts to give structure and momentum to the implementation phase. It is also a useful way of monitoring who has responsibility for different tasks. Action planning can be broken down into six elements:

(a) Identify (so far as possible) the duration or final deadline for implementation.

(b) Break implementation into coherent blocks of activity—each having some identifiable beginning and end point.

(c) Ascertain who will be responsible for completing each block—e.g., the client, or counsel, a colleague, or you? Is this a different person from the one managing implementation as a whole (if so this may beg the question of procedures for monitoring progress)?

(d) Make sure the goal(s) of each block are agreed and understood by those responsible for implementation.

(e) Attach a target date to each block.

(f) Note the resource requirements for each block.

An action plan can be presented very effectively in tabular form, and with varying degrees of detail depending on the complexity of the tasks involved. A typical work sheet is shown below.

Action planning work sheet

Objective:

Action	Responsible Person	Completion Date	Resource Needs	Monitored By
1.				
2.				
3.				
4.				
5.				
6.				

At its simplest it may be possible to present the whole implementation phase in one table, with each block listed sequentially under the 'action' heading. With more complex transactions it may be necessary to further sub-divide the process and prepare an action plan for each block. In this case the 'objective' would be the objective of the block, rather than the overall objective of the transaction, and the 'action' column would record the set of action steps which make up that particular block.

6.7.2 Contingency planning

We have already talked about the need to identify transactional risks and contingencies as part of Step 4 in our model, but these are also relevant to the task of implementation. Ideally a good solution will be relatively risk-sensitive, though not necessarily risk averse. At the risk of stating the obvious, there are plenty of things which can go wrong at the implementation stage. So, in addition to identifying the steps you should take to complete a case or transaction, it is also advisable to identify and, if possible, build in some allowance for those things that might go wrong. Basic contingency planning requires answers to three questions:

(a) What could go wrong?

(b) What could be done to prevent it?

(c) If it happens, how can we fix it?

Some contingencies, such as delay, could be built into the structure of your action plan; others might benefit from being recorded separately and held as notes on file. This is not least because a record of your preparedness to deal with contingencies can be a useful feature of your personal risk management system. This is the final topic to which we now turn.

6.7.3 Review and evaluation

In section **6.2.1** we made the point that the process model is cyclical, containing an implicit 'feedback loop' in which evaluation and implementation will often loop back into a further phase of problem (re-)identification. Even allowing for this cyclical process, all problems (hopefully) come to an end at some point. It is good—and increasingly common—practice to create a specific opportunity to review performance at this point (this could be just your own performance or a review of the whole team that you are part of). Review should be built into the process timeline at the outset, so that it is scheduled, and you know in advance that it will be triggered by a specific date or event.

6.8 Personal risk management

Very few claims and complaints against solicitors flow from the fact that they don't know or cannot apply the law. Most claims reflect either some sort of human error, or an absence of basic office systems or case work skills.

The bottom line is that it is probably impossible absolutely to prevent negligence (almost by definition), but it is possible to reduce the risks, not just of negligence, but simply of failing to deliver the quality of service that clients ought reasonably to expect. To do this, we must develop work practices that enable us to be aware of and to manage the risks that we will encounter. Much of the preceding chapter has, in fact, been implicitly about risk management. Our intention in this section is to draw together the connections between problem-solving and risk management in the form of a simple checklist. This should at least provide you with a starting point in developing your own risk management style. What follows constitutes a seven-step approach to personal risk management. We have cross-referenced this to the earlier sections of this chapter and other useful resources, as appropriate.

6.8.1 Step 1: identify the goals of the transaction

What is it you are trying to achieve? See **6.3.1**, **6.5.1** and **6.7.1** above.

6.8.2 Step 2: identify any risks likely to flow from the transaction

A 'risk' may be seen as anything that:

(a) impedes the attainment of those goals; or

(b) is an undesirable consequence of trying to attain those goals.

6.8.3 Step 3: consider grouping or categorising the risks identified

This may be helpful in complex transactions where there are multiple risks to be tracked. Possible groupings might include:

(a) *time management risks*—e.g., arising out of limitation of actions, or of deadlines agreed with the client/other side.

(b) *financial risks*—e.g., if the client is excessively exposed to financial loss by virtue of the action (whether as a consequence of the legal or other transaction costs), or where the action places the firm at some financial risk. The latter is particularly relevant if you are undertaking work on a conditional fee basis, in which case you should undertake a complete risk analysis before entering into a conditional fee arrangement: see O'Mahony, Marshall Ellson and Bennet, *Conditional Fees* (London: Sweet & Maxwell, 1999).

(c) *evidential risks*—e.g., because of gaps in the facts; uncertainty about the reliability of witnesses, etc.

(d) *legal risks*—i.e., those arising from uncertainty about the scope or operation of substantive or procedural rules, etc.

(e) *conduct risks*—be aware of your own exposure to legal or disciplinary action in respect of inadequate professional standards and misconduct. Key problem areas are likely to include: unreasonable delay in pursuing instructions; failure to keep clients informed of progress; acting where there are conflicts of interest, having no or inadequate authority to act (and hence creating potential exposure to actions for breach of warranty of authority); misleading the court, etc.

6.8.4 Step 4: identify/track the points in the transaction at which particular risks may arise

Risks may not be constant over the duration of a case or transaction. Good risk management flows from effective task and project management. See chiefly **6.3**, **6.4** and **6.5** above, and **Chapter 10**.

6.8.5 Step 5: characterise those risks identified as 'major' or 'minor'

According to criteria such as probability of occurrence, impact on transactional goals, ease with which the risk could be avoided.

6.8.6 Step 6: identify risk avoidance or limitation strategies for (at least) all major risks (if possible)

Identify specific steps you could take to minimise the particular risk(s) you have identified.

6.8.7 Step 7: agree an action plan

Identifying steps to be taken; personnel involved; priority and date for implementation, and, if appropriate, any review point—see **6.7.2**.

6.9 Concept check

As a concluding exercise, and a way of drawing the material together, you might like to check your understanding of the model we have developed:

(a) What is the primary rationale for adopting a process model of problem-solving?
(b) Whose goals and objectives drive problem-solving in legal practice?
(c) How might you distinguish between problem causes and effects?
(d) What are the five levels in 'five-level analysis'?
(e) What, according to Nathanson, are the two kinds of 'legal issue assessment'?
(f) Why is action planning important?
(g) What are the seven steps to effective risk management'?

Finally you might also find it helpful to re-visit the Written Standards presented at 6.2 and consider which techniques discussed in this chapter will assist you in fulfilling each 'problem-solving' Standard.

6.10 FURTHER READING

Crump, D., *How to Reason About the Law* (Newark, NJ: LexisNexis, 2001).

Freund, J., *Lawyering: A Realistic Approach to Legal Practice* (New York: Law Journal Seminars Press, 1977).

Hegland, K., *Trial and Practice Skills in a Nutshell*, 3rd ed. (St Paul, MN: West, 2002).

Kneeland, S., *Effective Problem-Solving: How to Understand the Process and Practise it Successfully* (Oxford: How To Books, 1999).

Levine, S., *Getting to Resolution: Turning Conflict into Collaboration* (San Francisco: Berrett-Koehler Publ., 1998).

McElhaney, J., 'The Trial Notebook' (1980) 1 *Litigation* 1.

Nathanson, S., *What Lawyers Do: A Problem-Solving Approach to Legal Practice* (London: Sweet & Maxwell, 1997).

O'Mahony, D., Marshall, D., Ellson, S., and Bennett, P., *Conditional Fees* (London: Sweet & Maxwell, 1999).

Pokras, S., *Successful Problem-Solving and Decision-Making*, 2nd ed. (London: Kogan Page, 1990).

Sherr, A., 'Lawyer and Client: The First Meeting' (1986) 49 *Modern Law Review* 323.

Sherr, A., *Legal Practice Handbook: Advocacy* (London: Blackstone Press, 1993).

Sherr, A., *Client Care for Lawyers* (London: Sweet & Maxwell, 1999).

Twining, W. & Miers, D., *How To Do Things With Rules*, 4th ed. (London: Butterworths, 1999).

7

Negotiation

7.1 Learning outcomes

This chapter deals in outline with the skills of legal negotiation. In this chapter we focus particularly on:

- the characteristics of effective negotiators;
- the skills of legal negotiation;
- negotiating style and strategy;
- common negotiation mistakes.

7.2 What is negotiation?

Negotiation is a process through which two or more parties arrive at a mutually satisfactory solution to a problem or dispute. The ability to negotiate effectively is an integral part of a lawyer's repertoire of professional skills. It is normally conducted with other lawyers who are representing other clients. In our experience students often find negotiation difficult. We don't quite know why this should be so, but we believe that the uncertain outcomes of negotiation may be one of the main reasons. For example, in the middle of a negotiation you may find that your plan doesn't work in the situation that has developed. You therefore have to adapt your strategy and behaviour to make progress. Moreover, in a negotiation the negotiator has to take full responsibility for the outcome, whereas with trial advocacy, for example, it is possible to shift some responsibility for the outcome on to the decision makers—judge, magistrates, jury. Furthermore, each negotiation develops its own boundaries and norms, so that the negotiators are responsible not only for the outcome, but also for the process.

Nevertheless, there are a number of characteristics common to all types of negotiation. If you understand these and keep them in mind, then you should be able to avoid most of the pitfalls and handle a variety of situations. Alan Fowler, writing on negotiation in the management context (*Negotiation: Skill and Strategies* (IPM, 1990)), suggests the following common characteristics:

(a) Negotiation involves two or more parties who need—or think they need—each other's involvement in achieving some desired outcome. There must be some common interest which puts or keeps the parties in contact.

(b) While sharing a degree of interest, the parties start with different objectives, and these differences initially prevent the achievement of an outcome.

(c) At least initially, the parties consider that negotiation is a more satisfactory way of trying to resolve their differences than alternatives such as litigation or arbitration.

(d) Each party considers that there is some possibility of persuading the other to modify their original position.

(e) Even when their ideal outcomes prove unattainable both parties retain some hope of an acceptable final agreement.

(f) Each party has some influence or power—real or assumed—over the other's ability to act. The power or influence may, however, be indirect and bear on issues other than those which are the direct subject of the negotiation.

(g) The negotiating process itself is one of interaction between people—in most cases by direct, verbal interchange. The progress of all types of negotiation is strongly influenced by emotion and attitudes, not just by the facts or logic of each party's arguments.

Bear in mind also that negotiating is a voluntary activity; either party can break away from or refuse to enter into discussion at any time.

7.3 Characteristics of effective negotiators

EXERCISE 7.1

List the characteristics of an effective negotiator.
If possible, compare your list with a colleague's.

To be a successful legal negotiator you need to develop a repertoire of *knowledge, skills and attitudes*. You need to *know* the context in which the negotiation occurs and the subject matter of the negotiation. You also need to *be able to*:

(a) analyse the legal issues, facts at issue and the parties' objectives and interests;
(b) listen, seek information, and give information;
(c) influence the other parties to the negotiation towards accepting your objectives;
(d) recognise when you have to concede and when you should stand your ground;
(e) recognise when you need to move on;
(f) think creatively; and
(g) review your negotiating experience.

You need to adopt appropriate *attitudes* towards:

(a) the process of negotiating;
(b) your client;
(c) the subject matter of the negotiation;
(d) the other party; and
(e) your own role in the process.

7.3.1 What do you need to know?

7.3.1.1 The context of the negotiation

Effective negotiators must be aware of the background and situational factors that surround each negotiation they take part in. They will therefore prepare themselves carefully by asking:

(a) Who is the client?

(b) Does the subject matter of the negotiation involve relationships which have to continue or which the parties want to continue; such as husband/wife, employer/employee, business partners?

(c) Or is it a 'one-shot' negotiation such as a personal injury claim?

(d) What is the client risking?

(e) How is their litigation funded?

(f) Who are the negotiators?

(g) Are they repeat players (lawyers who have expertise and experience in this type of litigation)?

(h) What do you know about how your opponent will approach the negotiation?

The answers to these questions will have a direct bearing on how you will approach the negotiation. Any legal negotiation will be shaped by the law and the possibility of court action if the negotiation fails is always present. You will find this referred to as 'bargaining in the shadow of the court'—if the parties can't reach a decision, then the judge will!

7.3.1.2 The subject matter of the negotiation

You must know the case file inside out. You need to have researched and have a detailed knowledge of the relevant law—remember it may not always be possible to go off to a library in the middle of a negotiation. Moreover, the other party will soon spot any possible advantage they can take because you are less well prepared than they are.

Thorough preparation will help you feel confident and in control.

7.3.2 What skills do you need?

7.3.2.1 Analysis of legal and factual issues, objectives and interests

Before you begin to negotiate you must assess:

(a) the factual and legal strengths and weaknesses of your case; and

(b) the factual and legal strengths and weaknesses of the other side's case.

Your fact analysis involves your identifying those facts which are agreed and those which are disputed. You will also have identified factual gaps and inconsistencies which you will need to clarify during the negotiation. Then you should apply the relevant law to the facts as alleged and assess the degree to which the law supports or undermines your case and the other party's case.

Your analysis of strengths and weaknesses will help you decide what it is you want to achieve. Work out the full range of potential objectives and prioritise them. Look for any underlying interests. For example, is a financial settlement the most important outcome for a client who has been unfairly dismissed, or are they really after reinstatement and an apology from their employer? Then identify and prioritise the other side's objectives. Are there any that both parties share? Take, for example, a breach of contract dispute: the parties have been doing business with each other for a number of years and would like to continue to do so. Finding a way to re-establish and maintain their relationship in the future may be more important to both of them than resolving this particular dispute. Their future relationship therefore becomes the central issue in the negotiation.

You must also be prepared for the negotiation to fail and plan what to do. This involves identifying your client's BATNA (Best Alternative to a Negotiated Agreement). This is the point at which you will walk away from the negotiation. A better alternative to agreement might be an agreement with another party, taking the dispute to mediation or arbitration, or proceeding to trial.

In summary, your plan should consider the following questions:

(a) What are the strengths and weaknesses of our/the other party's case?

(b) What are the central issues/the subsidiary issues/the peripheral issues?

(c) What is the central objective? What are the subsidiary aims and to what extent can these be traded in for the central objective?

(d) What are the main barriers to achieving what we want?

(e) What are the outside limits to an acceptable agreement?

(f) What information do we need to support our case?

(g) What are the arguments and information most likely to be used by the other party?

(h) What contingency plans are there if we fail to reach an acceptable agreement?

7.3.2.2 Listen, seek information, give information

When conducting a negotiation successful negotiators analyse the situation as it unfolds. This will involve your skills of active listening, reflecting back and questioning (See **Chapter 2**, 'Interviewing and Advising'). Your fact analysis will have identified what information you need from your opponent to fill gaps and clarify inconsistencies. However, the other side may try to conceal information from you. Use open and probing questions to seek information, and tune in to verbal responses and non-verbal communication to help you assess the openness or otherwise of your opponent's responses.

7.3.2.3 Influence the other parties to the negotiation towards accepting your objectives

This, too, is part of analysing the negotiation as it proceeds. You want to persuade your opponent to make concessions and move towards agreement on your terms. Your influencing skills are therefore vital in building understanding. Be positive and supportive in your language and non-verbal signals. Allow the other party to make their case fully before replying—never interrupt! Summarise at regular intervals to make sure you have understood each other correctly.

7.3.2.4 Recognise when you have to concede and when you should stand your ground

When you prepared your negotiation, you would have identified your 'bargaining counters', that is those things you would be prepared to concede if necessary in order to achieve more important objectives. The initial, exploratory phase of a negotiation is where each side explores the position of the other, and tries to discover what they will concede and what they will not.

The first rule on concessions is never to concede unless you get something you want in return. The second rule is only to concede those things that are relatively unimportant. The third rule is only to concede if there appears to be no real possibility of making progress otherwise.

7.3.2.5 Recognise when you need to move on

A negotiation can reach stalemate or deadlock for a variety of reasons. For example:

(a) The two parties appear to have widely differing objectives and interests.

(b) A party appears rigid (rather than firm) and refuses to make any concessions, even to keep the negotiation going.

(c) A party may attempt to coerce the opponent into making concessions.

Your client is paying you to negotiate an acceptable settlement, not to waste time going round in circles or scoring points off the opponent. You can agree not to agree on a particular issue for the time being and move on to the next point on your agenda. Alternatively, explore together what the barrier is to progress—you may find there is some underlying issue which the two of you are not addressing. If necessary, adjourn the meeting to allow time for reflection and consultation.

7.3.2.6 Think creatively

An effective negotiator thinks creatively and flexibly when both conducting and planning the negotiation. Negotiation is a dynamic process in which things may not necessarily go according to your plan. For example, your view of the case may change as more information is revealed; your opponent may come in with a style and strategy you had not anticipated; you may need to alter your strategy and tactics in response to a position taken up by the other side. A rigid structure is therefore not only unrealistic; it can also limit your chances of obtaining the outcome you want. Planning therefore involves much more than preparing the order in which you want to discuss matters; it involves thinking of as many ways as possible to resolve the issues, looking at the dispute from both sides and identifying possible solutions which could satisfy both parties' interests. It also means developing a repertoire of strategies to deal with a variety of approaches from your opponent. These will be built up from your experience as a negotiator, and depend on your ability to reflect and act on your experience.

7.3.2.7 Review your negotiating experience

After each negotiation you should take time to reflect on the process and outcomes. Ask yourself:

(a) Are you satisfied with the outcomes? Why/why not?

(b) What concessions were made? By whom?

(c) Were the concessions necessary ones?

(d) Who was the more effective negotiator? Why?

(e) What strategies and actions helped most?

(f) What actions hindered the progress of the negotiation?

(g) Did you trust the other party? Why/why not?

(h) What affected feelings most?

(i) How well was time used? Could it have been used better?

(j) How well did you listen to each other? Who talked more? Why?

(k) Were creative solutions suggested? What happened to them?

(l) Did you fully understand the other's underlying issues and concerns? Why/why not?

(m) How adequate was your preparation? How did this affect the negotiation?

(n) What were the strongest arguments put forward by the other party?

(o) How receptive was the other party to your arguments and ideas?

(p) What are your main learning points from this negotiation?

(q) What will you do differently next time?

(See Hiltrop and Udall, *The Essence of Negotiation* (London: Prentice Hall, 1995))

7.3.3 Develop appropriate attitudes towards:

7.3.3.1 The process of negotiating

Negotiations tend to follow a basic pattern. At the early stages the parties explore each other's position; they build a relationship and seek to understand each other's concerns. They identify the issues and familiarise themselves with the other's negotiating style. They then begin a process in which they begin to move towards each other. It is a process of exchange and persuasion. They make offers and respond to offers; they exchange information, seek to persuade the other side and each side's stance is modified by what is learned from the other. Gradually they begin to narrow their differences; they agree on some issues and identify fresh problems. They begin to see that agreement is at hand and move towards closure.

It is important to be aware of these processes and to plan accordingly. It is also important to be aware that different strategies may be appropriate at different phases in the negotiation process. Many negotiations start off competitively, trading bids and counter-bids, but as deadlock approaches there is a shift to a cooperative stance to reach agreement. It is important, too, to recognise that different strategies involve different sequences. A competitive bargainer in an adversarial negotiation typically exchanges information after trading bids. Bids will then be modified and concessions made in the light of the new information. A cooperative bargainer, on the other hand, seeks to exchange information, first identifying common interests, then working out solutions and finally moving to a process of bidding.

7.3.3.2 Your client

The vital thing to remember here is that it is the client's case. Your objectives will be based on what the client wants, not what you think they want, or what you think is best for them. Make sure you fully understand your client's objectives and interests, needs and priorities. As a competent professional, it is your responsibility to represent your client as effectively as you can and to support them in trying to achieve their objectives.

7.3.3.3 The subject matter of the negotiation

Similarly, you may find yourself out of sympathy with the subject matter of a negotiation. You may regard the issues as trivial, or they may offend your sense of right and wrong. Again, it is a part of your professional competence to represent your client's point of view.

7.3.3.4 The other party

You will meet a range of individuals and approaches during your negotiating career. Some of these will be difficult to deal with. Some may be aggressive, or rude, for example. In human relations, the behaviour of one party influences the behaviour of the other. Aggression invites aggression in response. Your professionalism demands that you don't get drawn into this kind of reciprocal behaviour. Remain detached and remember that you are there to reach an agreement if possible. Make this clear to your opponent.

7.3.3.5 Your own role in the process

In a negotiation you can act only with your client's authority. You must therefore be clear about the scope of your instructions and must keep your client fully informed to enable them, not you, to make decisions. You are under a duty to get the best possible settlement for your client. However, this duty can conflict with another aspect of your role. Lawyers are repeat players, in the sense that they will probably meet the same opponents regularly over time. They will therefore want to maintain good working relationships with their colleagues. If you gain a reputation as a very competitive negotiator, it could make it difficult when negotiating on behalf of future clients. On the other hand, maintaining good working relationships with fellow professionals should not be done at the expense of getting the best outcome for a client.

7.4 Negotiating styles and strategies

When planning for negotiation you need to be able to select a negotiating method which best suits the situation. Many law students, and some lawyers, rely on intuitive knowledge and skill when negotiating. Effective negotiators resist this temptation. They reflect on their negotiating experience and develop an operating theory or method to guide their practice. A negotiating theory or method is not made up of abstracted concepts. It consists instead of a set of working assumptions, accepted principles and rules of thumb which can be used to analyse, predict and understand the nature or behaviour of a certain situation (see Murray, Rau and Sherman, *Processes of Dispute Resolution* (Westbury: Foundation Press, 1988)). On the other hand, students and lawyers should not ignore theories of negotiation, since they distil the received wisdom of many other negotiators and have a lot to teach us.

7.4.1 Negotiating style

When we talk about negotiating style, we mean interpersonal behaviour—the way in which a negotiator communicates, verbally and non-verbally, with the other party. The negotiation literature identifies two distinct styles: competitive and cooperative. Competitive negotiators tend to be perceived as ambitious, egotistical, arrogant, clever, tough, dominant, forceful, aggressive and attacking. They aim to get the best settlement for their clients, but this also includes a reward for them—the satisfaction they gain from outdoing their opponents. They tend to view negotiation as a game, the purpose of which is to score as many points as possible and win a clear victory. Competitives make high opening demands, may use threats, stick to their opening position and be unwilling to reveal information to the other side.

These tactics are designed to undermine the opponent's confidence in the case so that he will agree to settle for less than he intended. However, these tactics are highly risky because if used against an equally assertive opponent, they can lead to deadlock. Furthermore, they can damage long-term relationships with other lawyers, which rely on mutual trust.

In contrast, cooperative negotiators are perceived as polite, friendly, conciliatory, tactful, sincere, careful and facilitating. Like competitors, they are concerned to obtain the best possible settlements for their clients but by a standard of fairness and ethical bargaining. They are willing to move from their opening position, share information, avoid threats and seek common ground. The cooperative style is believed to produce more favourable outcomes and result in fewer ultimate breakdowns in bargaining.

The major disadvantage of the cooperative style is its vulnerability to exploitation. If matched against a tough, non-cooperative opponent, the cooperative is thought to have an alarming tendency to continue discussing the case fairly and objectively, making concessions on the weaker aspects of their case. The competitive is able to accept all this cooperation without having to give anything in return. Competitives interpret cooperation as a sign of weakness, so from their point of view people who are strong or who have strong cases do not make concessions or admit weaknesses. When an opponent acts cooperatively with them, competitives increase their demands and expectations about what they will be able to obtain.

7.4.2 Negotiating strategy

Strategy is the general approach you take to getting what the client wants. The distinction between strategy and style is a subtle one and was developed by advocates of a problem solving approach to negotiation (Fisher and Ury, *Getting to Yes, Negotiating Agreement Without Giving In*; Menkel-Meadow *Towards Another View of Legal Negotiation: The Structure of Problem Solving*). The distinction was drawn as part of the case for a problem-solving approach. It was suggested that both the competitive and cooperative styles were informed by a similar adversarial strategy, and that problem-solving provided the only real effective alternative.

Both styles are seen to share a simple underpinning assumption: that each party wants as much as they can get from a fixed cake, so that one party's gain is the other party's loss. This is frequently referred to as a 'win/lose' orientation to negotiation. Parties structure the negotiation around a series of positions: opening position (the most you think you can get), target position (what you are happy to settle for), and bottom line (the least you will settle for). The parties engage in a stylised ritual of offer, response, counter-offer, counter-response.

A problem-solving, or 'win/win' approach has a different ethos. It seeks to uncover the underlying needs, interests and objectives of each party so as to open up greater opportunities for mutual gain. Lawyers negotiate with each other as participants in a joint decision-making process rather than as antagonists in a battle over the distribution of goods. A problem-solving strategy recognises the legitimacy of the values, positions and interests of the other party. It assumes that most negotiations involve multiple issues and that a strategy that focuses on underlying interests is most likely to uncover those issues. Negotiation becomes a means of establishing a positive relationship with the other party.

> The sale of a second-hand car has led to a dispute. Shortly after the sale, it ceased to function. The buyer (B) sues claiming consequential damages, including lost income because her employer deducted pay following repeated lateness and absences caused by the malfunctioning car. The seller (S) counterclaims for the balance due on the car.
>
> How might the dispute be resolved:
>
> (a) if an adversarial strategy is adopted?
>
> (b) if a problem-solving strategy is adopted?

If an adversarial strategy were adopted, the parties would structure their negotiations around the value of their respective money claims. B's lawyer would work out the amount her client had paid for the car, how much she has lost in income and the cost of the legal action. S's lawyer would subtract payments B has made from the purchase price and add the cost of recovering the balance. Both lawyers might then subtract from their targets the costs of achieving the results through court action. The solution to the dispute would be measured in monetary terms.

The disadvantage of this conventional adversarial approach is that the parties' real objectives may not be accomplished. B has to start all over again to find a reliable car. She is unlikely to return to this dealer. S has lost a customer as well as a sale. If the parties consider what they really wanted from the transaction, they might arrive at other solutions. If, for example, B's problem is transport to work S might repair her present car or substitute another at little or low cost. At the same time, S could continue to hold B to her contract, or a new contract could be negotiated (Menkel-Meadow, 'Towards another view of legal negotiation: the structure of problem-solving').

Which negotiating style and strategy you select will depend upon what is appropriate in the context, the goals of your client, the stage of the negotiation you have reached, and your personal preference. Negotiating style is a matter of personality as well as attitude, and you will want to use one which you feel happy with. It is important to realise, however, that you need to build up the ability to use a range of styles and strategies congruent with your values and personality, as well as to recognise when to use a particular one.

As we have said, the progress of negotiation is strongly influenced by the personal values, skills, perceptions, attitudes and emotions of the people at the bargaining table. It is one thing to set out theories and principles of negotiation and recommend you follow them, it is quite another to keep cool and stick to these principles when you are in the middle of a highly complex, potentially emotionally charged negotiating process. We therefore conclude this chapter on negotiation with an outline of some errors made by students and lawyers early in their negotiating careers.

7.5 Common negotiating mistakes

Hiltrop and Udall, writers on negotiation in management, suggest that negotiations which go wrong often do so for similar reasons. They call these reasons 'syndromes' and identify five of them. We summarise the four syndromes which we think inexperienced legal negotiators are most likely to fall victim to:

(a) *The 'one track' syndrome*. These negotiators have determined all the facts of the case and the solution they want in advance and enter the negotiation convinced that the other party will accept their version of the facts and their solution. Symptoms include beginning the negotiation with an agenda of key points to be covered in an apparently logical order. The points are then worked through quickly and mechanically, with no regard to the reactions of the other party, who will be prevented as far as possible from entering the 'discussion' and may be interrupted if they do.

Underlying this behaviour may be a lack of confidence in their own negotiating skills and an attempt to control the negotiation process. However, it's important to understand that negotiation is an interactive process involving joint decision-making about how to conduct it. An agenda planned in advance should be viewed only as a guide and need not be rigidly adhered to; the parties should agree a joint working plan for the discussions. Moreover, do not ignore the responses of the other party. Throughout the process effective negotiators listen actively and watch for non-verbal cues of agreement, dissent, frustration, etc.

(b) *The 'random walk' syndrome.* This can be paraphrased as getting nowhere, or going round in circles and still getting nowhere. Before a conclusion is reached on an issue, the negotiation jumps to another issue. Or the negotiation periodically returns to the same issue without adding anything to the discussion. This usually happens because the parties have not thought through the problem and all its ramifications before the meeting. Furthermore, parties may be unwilling to face up to conflicts which might arise if a problem area is probed in depth. The symptoms are failure to summarise the issues discussed and agreed. Alternatively, one negotiator attempts to summarise what they think the parties have agreed, but the other denies they have agreed.

The 'treatment' is clear—think the problem areas through before the negotiation and anticipate the areas of potential conflict. During the negotiation, obtain more information about the area under discussion, take time to define the problem more closely and seek common ground. Re-state and summarise positions regularly, to ensure both parties understand both points of view. Don't be afraid to build in a period of silence to allow time to think, and if necessary, adjourn the meeting to consider the discussion and reflect on the way forward.

(c) *The 'conflict avoidance' syndrome.* The parties do not address the real issues underlying their dispute. Instead they may concentrate on the less important areas of the dispute because these are more 'comfortable' to handle. Moreover, a party may be too eager to make concessions. The underlying causes may be that a negotiator is self-conscious and concerned about what people think of him and wants to come across as a very nice person. Alternatively, he wants the interaction to be as painless as possible and so identifies what the other party wants and is content to give it to them. This may occur because the negotiator feels pressured and intimidated by a competitive opponent.

The primary treatment for this syndrome is to focus on your function: why are you entering this negotiation? To get the best settlement you can for your client. Giving in is unlikely to enable you to do this! Identify the client's objectives and interests, prioritise them and plan concessions on an 'if'... 'then'... basis. During the negotiation, exchange information with the other side to identify and agree areas where agreement could be reached and how it could be reached.

(d) *The 'win-lose' syndrome.* The negotiators view, or come to view, the discussions as a contest or debate which they are determined to win. The main symptoms are:

(i) refusal to accept the validity of the other's views, claims and arguments;
(ii) unwillingness to reveal information or make concessions;
(iii) critical statements, personal attacks and emotional outbursts; and
(iv) closed questions and leading statements to gain compliance with preconceived ideas (for example, 'You must accept that... or 'You do not seem to be able to understand that...').

We discussed the potential danger of adopting a highly competitive style and strategy earlier in the chapter. Underlying this approach and behaviour seems to be the preconception that negotiating is a battle rather than a problem-solving forum. This may stem from the negotiator's perception that 'bargaining in the shadow of the court' involves the use of adversarial trial tactics. It is therefore important to resist the desire to reject or undermine every argument of the other party, even where you could agree. Ask yourself: why am I here? What are my client's interests? What are my own interests? How important is the longer-term relationship with the other party? How can I make them concede without losing face?

Finally, recognise that you are an experienced negotiator, even though you may not yet have participated in legal negotiation. For example, you may have bought or sold a car or a house, or you may be a parent.

> Based on what you have just read, your negotiating experience and reflections, what improvement in your knowledge, skills or attitudes would you most like to achieve?
> How will you set about doing this?

7.6 Learning outcomes

In this chapter we looked at the skills, knowledge and attitudes needed to develop your capacity to negotiate successfully.

You should now be able to:

- describe the characteristics of effective negotiators;
- list the skills needed to negotiate successfully;
- evaluate and select an appropriate negotiating style and strategy;
- represent your client's interests effectively in a legal negotiation;
- reflect on and develop your negotiating skills, knowledge and attitudes.

7.7 Self-test questions

(1) What characteristics do all types of negotiation have in common?

(2) To be a successful negotiator you need to develop a repertoire of knowledge, skills and attitudes. List these.

(3) What is meant by the terms 'negotiating style' and 'negotiating strategy'?

(4) What are the four 'syndromes' most likely to derail a legal negotiation?

7.8 FURTHER READING

Fisher, F. R., Ury, W., and Patton, B., *Getting to Yes: Negotiating Agreement Without Giving In* (London: Arrow, 1997).

Fowler, A., *Negotiation: Skills and Strategies* (London: IPM, 1990).

Halpern, A., *Negotiating Skills* (London: Blackstone Press, 1992).

Hiltrop, J. M., & Udall, S., *The Essence of Negotiation* (London: Prentice Hall, 1995).

Menkel-Meadow, C., 'Towards another view of legal negotiation: the structure of problem-solving' (1984) 31 *UCLA L Rev* 754.

Murray, J. S., Rau, A. S., and Sherman, E. F., *Processes of Dispute Resolution* (Westbury: Foundation Press, 1988).

8

Advocacy and the solicitor

8.1 Learning outcomes

This chapter deals with the skills of advocacy. In this chapter we focus particularly on:

- identifying key principles of the ethics of advocacy;
- understanding the structure of a case presented for trial;
- developing a case presentation strategy incorporating the client's goals;
- making an effective submission;
- planning an examination-in-chief and cross-examination using appropriate questions.

8.2 Advocacy

Advocacy, in the common law tradition, is associated with the presentation to a judge of two competing versions of the facts, of the law, or both. It is the activity most clearly associated with public images of the lawyer in the media. Despite increasing emphasis on written evidence and the growth of mediation, it is unlikely that the supremacy of adversarial advocacy will be supplanted. With improved opportunities for solicitor advocacy its significance for solicitors may even increase. Trainees and newly qualified solicitors generally get a taste of advocacy by appearing in court summonses, directions hearings, etc. But, even if advocacy does not interest you, understanding it provides vital insights into case preparation.

8.3 Solicitors' rights of audience

Under s. 36, Access to Justice Act 1999 solicitors share with barristers a right of audience before every court. In the case of solicitors, however, the right to appear as an advocate in higher courts is subject to satisfying the requirements of the Law Society's qualification regime. This is currently established under the Courts and Legal Services Act 1990 as amended by the Administration of Justice Act 1999. Solicitors have automatic rights of audience, although in some cases limited, in the following four courts:

(a) Magistrates' courts: crime, care proceedings in respect of juveniles, licensing applications and a sizeable domestic jurisdiction;

(b) Crown Court: appeals to the Crown Court from a magistrates' court or on committal for sentence, if the solicitor appeared in the court below, or where the Crown Court hears appeals from the civil jurisdiction of the magistrates' court; full rights of audience in some provincial Crown Courts (because they had such rights before the quarter sessions became the Crown Court), e.g., Carnarvon, Barnstable, Truro, Doncaster and parts of Lincoln;

(c) County courts: full rights of audience in the county courts covering, inter alia, debt, breach of contract, personal injury, recovery of land, bankruptcy and insolvency, admiralty, family proceedings and miscellaneous matters such as mental health and sex discrimination;

(d) The High Court: no general rights of audience, but a right to appear when the judge or master sits in chambers. Solicitors may also appear in bankruptcy matters, in formal or unopposed matters where the court will not be called upon to exercise any discretion and upon judgment in open court following a hearing in chambers in which that solicitor appeared;

8.4 The Law Society's training requirements

The Law Society aims to lay a solid foundation for advocacy with a training regime in four phases. The first three are compulsory and the last applies only to solicitors who wish to take advantage of full rights of audience. The first phase is the Legal Practice Course. On completing the course you should be able to formulate a coherent submission based upon facts, general principles and legal authority in a structured, concise and persuasive manner. You should understand the crucial importance of preparation and the best way to undertake it. You should be able to demonstrate an understanding of the basic skills in the presentation of cases before various courts and tribunals and should be able to:

(1) identify the client's goals;

(2) deal with appropriate client care and professional conduct issues in preparing and presenting the case;

(3) identify and analyse factual material;

(4) identify the legal context in which factual issues arise;

(5) relate the central legal and factual issues to each other;

(6) state in summary form the strengths and weaknesses of the case from each party's perspective;

(7) develop a case presentation strategy;

(8) outline the facts in simple narrative form;

(9) prepare in simple form the legal framework of the case;

(10) prepare the submission as a series of propositions based on the evidence;

(11) identify, analyse and assess the specific communication skills and techniques employed by the presenting advocate;

(12) demonstrate an understanding of the ethics, etiquette and conventions of advocacy.

The Legal Practice Course introduces the general principles involved in advocacy through role-play and simulation. In Civil Litigation and Criminal Litigation you are given instruction on the appropriate pre-trial procedures.

During the Professional Skills Course, which is taken during the training contract, you will be expected to advise clients on pre-trial proceedings, understand the role of preparation and how to undertake it, and prepare and assist in pre-trial procedures and proceedings. Trainees should be able to make a simple interlocutory application and, in the context of a civil and a criminal case to:

- use language appropriate to the client, witness(es) and triers of fact and law;
- listen, observe and interpret the behaviour of triers of fact and law, clients, witness(es), and other advocates and be able to respond to this behaviour as appropriate;
- speak and question effectively and thereby competently use appropriate presentation skills to open and close a case;
- use a variety of questioning skills to conduct examination in chief, cross-examination and re-examination;
- prepare and present a coherent submission based upon facts, general principles and legal authority in a structured, concise, and persuasive manner;
- present a submission as a series of propositions based on the evidence;
- organise and present evidence in a coherent form.

Trainees should also be able to identify and act upon the ethical problems that arise in the course of a trial.

During the training contract you should be given practical opportunities that will enable you to understand the principles involved in preparing, conducting and presenting a case. You could be asked:

(a) to help advise on pre-trial procedures;

(b) to help prepare cases before trial;

(c) in the company of one or more lawyers to:

  (i) attend the magistrates' court to observe trials, bail applications, pleas of mitigation or committal;

  (ii) observe the conduct of a submission in chambers or examination, cross-examination and re-examination in open court;

(d) to observe proceedings in family cases, industrial tribunals, planning tribunals or other statutory tribunals or the use of alternative forums of dispute resolution; or

(e) as training progresses, and under appropriate supervision, to take a more active role in the conduct of a case: this could include interlocutory applications before a master or district judge.

Finally, after gaining experience as a solicitor in lower courts and tribunals, you can apply for one of three higher courts qualifications. For this you must currently pass written and practical examinations (see the Higher Court Qualification Regulations 2000 and *The Guide to the Professional Conduct of Solicitors*, Ch. 2, para. 07 or rule 11 of the new Law Society's Code of Conduct).

Other Legal Practice Guides deal with the specific requirements of procedure and evidence that underpin the work of an advocate (see Craig Osborne, *Civil Litigation* and Hannibal and Mountford, *Criminal Litigation Handbook*). Here we are concerned to introduce the skills that tend to be applicable in any jurisdiction and to any trial, whether civil or criminal and whether before judge or jury.

8.5 The skills of the advocate

This text is concerned with competence, which comes with learning and practising the skills of advocacy. How effective an advocate you can be will generally only be known after you have acquired many years' experience. As a beginner it is important to get the basic presentation skills right. Remember, the point of advocacy is persuasion. Be guided by what is, and what is not, persuasive and bear the following points in mind:

- Never merely read a prepared text because the presentation will lack authenticity, which is not persuasive.
- Be confident and communicate, holding the audience's attention.
- Keep your head up so that you can maintain eye contact with, and address, the audience.
- Do not rush; speak clearly and slowly.
- Do not move around excessively, gesticulate, scratch or do anything else that may distract the audience from your message.
- In most matters, even complex ones, it is best to make a few simple propositions the essence of your case. Select them, articulate them clearly, reinforce them.

(See further Boon, *Advocacy*, Chapter 1, 8.10.)

The starting point of most advocacy experience is a simple submission. This may occur pre- or post-trial in a summons, bail application or plea in mitigation. Much of what follows in this chapter concerning preparation (see **8.7**) and speeches (see **8.9**) is relevant to submissions. A distinctive feature of submissions, compared to speeches, however, is the likelihood of interventions by the person hearing the submission. He or she may wish to clarify the facts, understand the argument, or test its validity. It is important to do everything possible to leave the arbiter satisfied that their point has been dealt with. Never rebuff the arbiter by offering to return to the point later. Without being obsequious or fawning, endeavour to establish a polite and trusting relationship with the arbiter. If the arbiter is in error, do not suggest as much. Correct the error diplomatically, even apologetically, rephrasing the point as if the error is your own, for example, 'Master, I do apologise. Let me put the point another way.' (See further Nathanson, 8.10.)

8.6 Etiquette, ethics and conduct

8.6.1 Fundamental principles

Two fundamental principles underpin much advocacy practice:

You should not give your own opinion of the facts or the law. You construct the theory of the case, the story, based on the facts. Elicit the witnesses' testimony in a way that supports your theory but do not say, for example, 'It is obvious that the car was going too fast under the existing conditions' or 'I happen to know that piece of road is dangerous in icy conditions'. These principles underpin the prohibition on leading questions during examination-in-chief and the convention that advocates 'submit' and 'suggest'. But also, human nature being what it is, it is better to lead fact-finders gently than to insist that they agree with you (see further, Evans, **8.10**).

8.6.2 Ethics and conduct

The ethics of advocacy are more complex than other legal roles. They are summarised as follows by Lord Hoffman in *Hall* v *Simons*:

> Lawyers conducting litigation owe a divided loyalty. They have a duty to their clients, but they may not win by whatever means. They also owe a duty to the court and the administration of justice. They may not mislead the court or allow the judge to take what they know to be a bad point in their favour. They must cite all relevant law, whether for or against their case. They may not make imputations of dishonesty unless they have been given the information to support them. They should not waste time on irrelevancies even if the client thinks they are important. Sometimes the performance of these duties to the court may annoy the client. ...
>
> (*Arthur J S Hall & Co (a firm)* v *Simons, Barratt* v *Ansell and others* (trading as Woolf Seddon (a firm), *Harris* v *Scholfield Roberts & Hill (a firm) and another* [2000] 3 All ER 673 at 687g–h)

Particularly in relation to trial work it is important to explore the nature of the conflict between the duty to the client and to the court in more detail (see further Boon and Levin, Ch. 14, 8.10). The remainder of this section deals with the most basic points of etiquette. The Access to Justice Act 1999 underlines the duty of any advocate appearing before a court to act with independence in the interest of justice and to comply with the rules of conduct of the authorised body granting the rights of audience (s. 42(2A)). The Law Society's Code for Advocacy is set out in Annex 21A in The Guide to the Professional Conduct of Solicitors, 'The Guide' (and see Annexes 21Bto 21H). Although not yet in force, the new draft Law Society Code of Conduct deals with Advocacy in rule 11. In cases where the code really does not help the Professional Ethics Division of the Law Society will give advice or suggest alternative sources of help (see The Guide para. 1.09).

In addition to knowing the Code, you need to explain the conventions of the court or tribunal to your client and witnesses. The next section deals only with the basics: the formalities involved in addressing the bench, other advocates and witnesses; and how to deal with witnesses.

8.6.2.1 Addressing the bench

You should address members of the bench in the following ways:

(a) Magistrates should be addressed as 'Sir' or 'Madam'. It is not correct to call a magistrate 'your worship', although you may use 'your worships' when addressing the full bench.

(b) Circuit judges are referred to as 'Your honour'.

(c) The district judge of the High Court and the county court registrar should be addressed as 'Sir' or 'Madam'.

(d) A master of the Supreme Court, or a taxing master, should be addressed as 'Master'.

8.6.2.2 Addressing the other side

You should refer to 'My learned friend' only if the other side is represented by a barrister. Where the other side is represented by a solicitor, the convention is to refer to them as 'My friend'. A litigant in person should be referred to as 'The claimant/defendant', or 'Mr/Miss/Mrs/Ms Brown'.

8.6.2.3 Addressing witnesses

Witnesses should be addressed directly: Mr, Mrs, Miss, Ms, Dr. Except in the case of children, witnesses should not be addressed by their forenames.

8.6.3 The place of witnesses

If the defendant in a criminal case is to give evidence, he or she must be called before other defence witnesses. If the claimant in a civil case is to give evidence, he or she should be called before other witnesses. If the defendant in a civil case is to give evidence, he or she should be called before other defence witnesses. In criminal cases all witnesses, other than the defendant, should be kept out of court until called. In a civil case witnesses need not be kept out of court, but if you want them excluded you will need to give reasons to the judge. The object of excluding witnesses from the court is to prevent them from knowing what evidence is being elicited from others. This principle should be explained to witnesses, who may nevertheless stay in court once they have delivered their evidence. Witnesses should not discuss their evidence with others still waiting, nor should they discuss it with you if the case has been adjourned while they are giving evidence. You should always arrive at the court with sufficient time to explain the physical set-up and process to your client and witnesses and to help them feel comfortable.

8.6.4 Introductions

If opening a case you should always introduce yourself and the other side to the court, for example:

> May it please you (Your Honour, Madam/Sir, Master), in this case I appear for the prosecution/
> claimant and Mrs Smith (or my learned friend, if the advocate for the other side is a member of the Bar) appears for the defence.

You should arrive at court in plenty of time to fill in a slip stating who you are. If you have not had time to do this, after explaining whom you appear for, you should add your name.

8.7 Preparation

Preparation, not eloquence, is the key to successful advocacy because thorough preparation gives you the confidence to make an effective presentation. You need to review the file and be completely on top of the facts, the law and the procedural steps taken to date. Your strategy will be dictated by the formal case papers and your judgements about their interaction with the law. You will need a thorough understanding of the statement of case, or charges, and you will have to review the probable testimony of all anticipated witnesses. The next two sections consider developing a case strategy through the 'theory of the case' and the 'story' behind the case. This and other key documents should be organised in a case file.

8.7.1 The 'theory of the case'

The 'theory of the case' seeks to reconcile all the undisputed and disputed evidence to be presented at trial. It integrates the agreed facts with your version of the disputed facts to create a cohesive, consistent, logical and persuasive position. Your theory should guide you in formulating your argument, deciding what evidence to call and what evidence of the other side must be attacked or undermined. Consider the following examples:

(a) In a murder case, the prosecution's evidence will show that, following an argument, the victim was shot by a man some witnesses will identify as the accused. As advocate for the defence, two theories deserve consideration:
  (i) the accused did not do the shooting (wrong identification); or
  (ii) the accused did do the shooting, but was using reasonable force in his own defence.

(b) In a 'running down' case, the claimant pedestrian was struck by the defendant's car at a crossroads. Some evidence will place the claimant within a zebra crossing with the traffic lights providing a green signal for her to walk. Other testimony will show the claimant was outside the zebra crossing.

As the claimant's advocate, how many possible theories of the case can you identify?

These are simple illustrations of developing a theory of the case. You choose the option, a theory that is the best 'fit' with the facts, the one that is most robust. This provides you with an orientation towards the facts; it provides you with a consistent and logical position throughout the case. Your decision affects how you examine and cross-examine witnesses and provides the structure to your opening and closing arguments. Failure to develop a theory leads to a general lack of direction. Most important, a theory of the case enables you to identify the focal points in a trial, the admissibility of an exhibit, or the impression made by a witness, which may be missed if you have no overall plan. This is becoming increasingly important with the use of skeleton arguments before trial (see *Practice Direction Civil litigation: Case Management* (1995) All ER 385) because, ideally, your strategy should reflect your skeleton argument.

To develop a theory of the case you need to take several analytical steps. For example, where you are acting for the claimant, there are four steps to developing a theory, as illustrated in the diagram on the following page.

Develop as much additional evidence as you can to support your theory of the case, as well as to attack your opponent's. If your case rests on an alibi, you must search for facts that make that alibi as credible as possible. If an important piece of evidence consists of hearsay, you must find a basis for an exception to the hearsay rule that makes it admissible.

Your theory of the case is the most credible account of 'how it really happened'. It changes an amorphous search for truth into a specific line of argument that you can build upon. But, as Paul Bergman points out, the formation of a theory is not without risks. If you lock into a particular theory of the case too early, you may discard alternative and stronger theories and ignore inconsistent leads. As a safeguard, develop and test alternative theories and keep these in play. You can then adopt a flexible approach both to trial planning and to the trial itself. It is also important to be aware of ethical considerations. Do not allow your 'theory' to lose touch with the client's account or use it to shape the client's evidence or that of other witnesses. Remember the advocate's overriding duty to the court and to justice.

Cause of action

Review the elements of each cause of action you have brought in the case and the applicable law.

↓

Proof

Analyse how you intend to prove each of the required elements through available witness and exhibits.

↓

Areas of dispute

Analyse the contradictory facts that your opponent has available to determine what facts will be disputed, and anticipate the witnesses and exhibits your opponent will probably present in order to put those facts in issue.

↓

Evidence

Research all the possible evidentiary problems that arise so that you will be able to maximise the admissiblity of your proof and minimise that of your opponents.

↓

Strengths and weaknesses

Review the admissible evidence you and your opponent have on each element of required proof to determine the greatest weaknesses on both sides.

↓

Theory

Evolve an explanation that fits the legal and factual background and works the evidence into a coherent and credible whole.

Once your theory of the case is developed you should perform the same analysis to determine what your opponent's probable theory and position on the disputed facts will be. This is important preparation for cross-examination (see **8.8.7**).

8.7.2 What's the story?

The narrative element of a case is important. A case is presented primarily through the testimony of witnesses and the presentation of real evidence, exhibits and documents. Narrative works through *implication*, a form of argument which does not appear as an argument. It proceeds through a series of propositions, which accumulate, and gradually converge, leading to the inevitable conclusion.

The essence of narrative is characters, events and descriptions of place and mood. The characters display feelings, take part in dialogue, and engage in activities. A story will be more or less credible depending on the manner in which these details are sketched. The story, and the various themes that make up the story, will emerge not in the form of a simple statement, but in the form of the connections drawn between each of the elements within the narrative.

The concept of trial arguments as narrative structures was given empirical foundation in Bennett and Feldman's analysis of trial strategies in American courts (Bennett and Feldman, *Reconstructing Reality in the Courtroom* (London: Tavistock, 1981)). A narrative links five elements: the scene, the act, the agent, the agency, and the purpose. This is most clearly demonstrated in a criminal case where the judge's instructions to the jury emphasise the prosecution's obligation to construct a structurally complete story:

In order to convict the defendant of a crime, for example, the prosecution must prove beyond reasonable doubt that:

(Actor)	The said defendant, . . .
(Scene)	did on the charged date/time/place/occasion etc.,
(Purpose)	wilfully/knowingly with intent etc.,
(Agency)	use force/cause the victim fear of bodily harm/offer for sale etc.,
(Act)	to take the property of/to cause the death of etc.

EXERCISE 8.1

Pick any civil cause of action and identify the five narrative elements.

The basic prosecution or claimant task is to represent the defendant's action within a coherent set of scenes, agencies and purposes as the action develops over time. To satisfy the minimum structural criteria of a story you must first *situate* the actor and action in time and space, at the scene; second, establish the actor's purpose or *intent*; third, consider the behavioural mechanics or *execution* of the act. Establishing a story provides consistent definitions for actors and acts that are constant across the scenes, agencies and purposes, thus yielding a consistent and clear interpretation for the act.

While the prosecution or claimant have to construct a whole story, offering its own complete story is only one option for the defence. The defence also has the option of *challenging* the prosecution's story. It can be argued that there are missing elements in the prosecution's story, or that definitions of scenes, acts, actors, agents or purposes do not support the same interpretation of the defendant's behaviour. It can *redefine* the story, showing that a different meaning emerges when slight changes are made in the interpretation of the evidence. Finally, the defence can *reconstruct* the evidence, and tell the defendant's own story. Here the defendant's advocate uses the evidence to tell a completely different story about the defendant's behaviour.

Constructing the case in the form of a story enables you to make tactical decisions about how to place evidence, order witnesses, structure questions and present arguments. The concept of a story can be used to develop the argument in a case through the manner in which it informs three key elements in the case:

(a) *Definitions*—facts are constructed into a story line through the specific language used by witnesses (and elicited by lawyers) to define pieces of evidence.

(b) *Connections*—by locating or placing a particular piece of evidence in relation to the other elements in the story, connections are established for the arbiter.

(c) *Validations*—the integrity of a story depends on whether the definitions and connections can be validated through supporting definitions or invalidated by alternative definitions and connections that are equally plausible.

The 'story' develops the theory of the case into a particular narrative and suggests the tactical decisions made during a trial. Deciding tactics involves two sets of decisions. First, which witnesses to call, and in what order? Second, how will you make your story plausible? Issues of plausibility or credibility are considered next while the selection and ordering of witnesses is considered in **8.8.3.1**.

8.7.3 Credibility

To establish and enhance the credibility of your case, ensure that your evidence is consistent with common sense, consistent within itself, and consistent with the established facts (see Bergman, *Trial Advocacy in a Nutshell*, 8.10).

8.7.3.1 Is the evidence consistent with common sense?

Evidence will not be credible if it is counter-intuitive. Thus, if you wish to introduce a witness who claims to identify X, whom they saw running away from them a hundred yards away, on a dark, badly lit night, you will need to introduce evidence to show that claim is credible. Similarly, if your client claims that he had left his house at 3.00 a.m. to return a book to the library, you will need to introduce evidence to make it plausible. We know from our own experience that these claims are unlikely.

Consider whether the definitions of behaviour that you are making, or the connections that you are making, are credible. Do they make intuitive sense? If not, you need to consider introducing validating evidence. Consider the following testimony:

Q1. At what point did Mrs Brown step out?

A1. Twenty yards beyond the zebra crossing.

Q2. Where was the bus at this point?

A2. Approaching the zebra crossing.

Q3. Where was the defendant's van?

A3. In the outside lane passing the bus.

Q4. What happened next?

A4. Mrs X crashed into the defendant's van.

What issue of plausibility is raised by the answer and how might it be resolved?

To make this answer plausible you need to ask other questions, to make it consistent with common experience. So evidence that Mrs A was late for work, that she was running along the pavement before running into the road, that she was distracted by a friend calling on the other side of the road would make *connections* between the statement and common experience. Note, too, that by moving from closed questions to an open question the advocate is able to *validate* the evidence by having the witness volunteer the phrasing.

8.7.3.2 Is the evidence consistent with itself?

People who give inconsistent evidence cast doubt on the remainder of their evidence. Consider the following example. George Kelly is identified by a witness as the person at the scene of the crime. The witness says that George was wearing a red T-shirt. George then gives evidence. In response to questions he denies being present and that it could not have been him; he does not have a red T-shirt. The prosecution then introduce a photograph of him wearing a red T-shirt. Not only does this cast doubt on his statement, it also undermines the whole of his testimony.

In preparing your case you need to search the expected testimony for inconsistencies and capitalise on them or, if they are working against you, de-emphasise them. You also need to determine the persuasive effect of inconsistent statements. Is the point crucial or relatively minor? How great is the inconsistency? Does the witness have a good explanation for the inconsistency? To what extent does the inconsistency at one point infect the rest of the witness's testimony?

8.7.3.3 Is the testimony consistent with established facts?

In developing a theory of the case you will seek to integrate disputed facts with agreed facts to produce a consistent and coherent argument. When questioning witnesses you should seek to establish agreed facts and avoid inconsistencies. If you have agreed with the other side that your client was wearing a red T-shirt, and that a person with a red T-shirt was identified as the thief, there is no point in one of your witnesses saying that your client was wearing a white T-shirt. The witness appears to be making up evidence to assist the defence case. The lesson is that in preparing your questions you need to avoid disagreements on agreed facts and make *connections* between the agreed facts and disputed facts.

8.7.4 Applications and overview

8.7.4.1 Planning

Sir David Napley (*The Technique of Persuasion*—see **8.10**) describes the development of a plan. He urges advocates to review:

(a) the object to be attained;

(b) the factors which affect the attainment of the object;

(c) courses of action open to you and the other side; and

(d) the plan.

Napley illustrates his approach using the case of a Nigerian lady who has been charged with shoplifting.

She says she had gone to a large chemist's to purchase presents for her return to Nigeria. She had often shopped at the store, spending £120 that morning. She is the wife of a government official and was scheduled to leave at 2 p.m. that afternoon. She was in a great hurry; she had to visit a sick relative, make other purchases, change a considerable quantity of money, collect her luggage and get to the airport by 1 p.m.

The store was busy. She selected some ornaments from the first floor, worth about £13, and queued to pay. Because of the queue, she decided to pay downstairs. She saw there was a queue around this till, too, and so decided to save time by crossing the road to a bank, to change her money, with every intention of returning to pay. As she reached the pavement, a store detective, who then called the police, challenged her. She explained to the police officer that she intended going to the bank before returning to the store. On further questioning you find that because she was in such a hurry she had arranged with a car hire service to have a taxi waiting for her at 12 noon at the back of the store. She was arrested at 11.50 a.m.

EXERCISE 8.2

As representative of the accused, devise a defence plan.

8.7.4.2 The object to be attained and the factors which affect the attainment of the object

Your primary object is to secure an acquittal. A number of factors affect the attainment of that object. These are:

(a) The vast majority of shoplifting cases are properly charged and convicted. Magistrates tend to become case-hardened and store detectives know what the magistrates expect to hear. Your task is to convince them this case is the exception rather than the rule.

(b) The store detective will be an experienced witness and you have no material with which to refute her evidence. But...

(c) Your client had just enough money on her to pay for her purchases and a great deal more in dollars. But this is common in many shoplifting cases and may not help.

(d) The sole issue is whether the client was leaving the shop having stolen the articles, or whether she intended to return and pay.

The following factors need to be taken into account:

Against her	For her
(a) She did not at once say she was going to the bank and returning to pay.	(a) She had a taxi waiting at the rear entrance. If she was making a getaway she would have been more likely to head for her waiting taxi.
(b) She knew she had not paid for the goods when she left the shop.	(b) When stopped she was about to cross the road in the direction of the bank.
(c) She told no one before leaving what she intended to do.	(c) She told the store detective and the police officer, on the latter's arrival, what had been her intentions.
	(d) She was of excellent character and standing.
	(e) The bank confirms no one has inquired about whether she was known there and changed currency there.

The next stage of the analysis is to decide the course of action open to the defence and to the prosecution. The defence must call the taxi driver to prove he was waiting for the accused at the rear of the shop. Two strong and impressive character witnesses should be available. The shop must be viewed in advance of the hearing to familiarise yourself with the layout. If possible, someone from the bank should be called to prove she was known to change currency there. The prosecutor should be asked whether there are any facts unknown to you.

8.7.4.3 The plan

There is no real prospect of persuading the magistrates that your client is innocent. Your objective, therefore, is to persuade them that there is reasonable doubt about her guilt, which means that they must acquit. Your tactics could be to impress the bench into accepting that your client acted foolishly and not wickedly, and that she was proceeding in the direction of the bank and not in the direction of the taxi. Cross-examination and the argument should be limited to that one issue. As the case develops, however, you get a bonus. The police officer describes how he told the accused she had committed a crime by leaving the shop with the goods. It can now be suggested that she had been charged without regard to the need to establish dishonest intent.

> What questions do you ask the store detective?

Remember to lay a proper foundation for your questioning. First ask the store detective whether there is a Barclays Bank opposite the shop. You then ask whether she was present when the police interviewed the defendant. You will ask whether your client had said that she was going to the bank to cash some money. You might then ask her whether she had been to the bank to see whether the defendant was known there. Note that you already know the answers to all these questions. By asking them you are establishing the platform for the next and crucial stage.

You must take care in phrasing your next question. Your next step is to have the prosecution witness concede the possibility that the client's account was a true one. If you simply ask her if she agrees that your client was going in the direction of the bank, she will say that she has no idea. But if you put the question in this form: 'You of course are not able to dispute the possibility that this lady was intending to go to the bank?' you should get a positive answer.

8.7.5 Summary

It is important to plan advocacy. There are different approaches to planning but they all have similar objectives: presenting a coherent and engaging case, identifying the key issues and clarifying how, as a matter of tactics, these key elements of the case should be developed through the presentation of evidence and in speeches to the court.

8.8 Examining witnesses

8.8.1 Examination-in-chief

An examination-in-chief is the process in which you elicit evidence from your own witnesses. It is a process that to some extent is being overtaken by the Civil Procedure Rules, which suggest that written statements, witness statements or affidavits, should ordinarily stand as evidence-in-chief of each witness (CPR Part 32.5(2)). It is necessary to understand the principles of establishing evidence in chief for the purpose of preparing written and oral evidence. The court may also allow 'amplification' of a witness statement by oral evidence (CPR Part 32.5(4)). The idea is to develop through witnesses' answers a case that is clear, logical and forceful.

It is important that you ensure that the witness gives clear evidence at a speed which allows a note to be taken. It might be necessary to ask the witness politely to speak up and face the judge or to slow down. You might say something like, 'Mr Brown, could I ask you to give your evidence more slowly? I suggest that you watch His Honour's pen'.

8.8.2 General principles

The conduct of an examination-in-chief presents a dilemma. If you provide too much detail the arbiter, whether professional judge or jury, may be overwhelmed and bored. If you provide insufficient detail you may not establish your case. You resolve this dilemma by presenting only that evidence which is relevant, by organising it logically, and by controlling the pace at which it is heard.

8.8.2.1 Relevance

The arbiter wants to hear evidence that is relevant. Ask yourself:

(a) Does a line of questioning establish a point important enough to be included in your closing argument?

(b) If you were in your opponent's shoes, is it the kind of evidence you would bother to contest?

(c) Is there a direct link between the testimony and the desired conclusion?

If the answer is 'yes' to each question, the evidence is relevant. This does not mean that all of the material should be excluded. It may provide important background or flavour. But always bear in mind the reason for particular questions.

8.8.2.2 A logical structure

A chronological presentation of the evidence is usually best, particularly if you are using a narrative structure. A chronological narrative should stimulate a witness's memory to provide a fuller and more credible story and stimulate the fact-finder's memory, so that they are more likely to remember the evidence. This is not a universal rule. It may be best to start with the most interesting thing a witness has to say which may have happened in the middle of the chronological sequence. Start the examination at that point, then backtrack and proceed through the rest of the testimony in chronological order, finishing on the same strong point.

8.8.2.3 Use language and pace to vary your approach

When you phrase your questions you should choose simple words and phrases and choose them carefully. If you ask a witness 'how fast' a car was moving, you are likely to get a different answer than if you had asked 'how slow' it was moving.

Use questions to control the testimony. Carefully directed questions produce a detailed account in a series of short answers, slowing the action right down. Using broad open questions produces a very different response, a quick account of the action, which can be useful in conveying the speed at which it developed.

8.8.2.4 Use points of reference and transitions

Points of reference and transition questions can be used to change pace and structure the examination. A point of reference is a matter testified to by a witness included to provide a context for a later question. The third question in the following sequence includes a point of reference:

(a) What happened next?

(b) Please go on.

(c) What happened directly after the car swerved?

Points of reference can be used to emphasise important facts, clarify confusing testimony, obtain greater detail, or provide a transition from one episode to another. They can add variety, even drama, to your questioning. Transition questions are used as signposts. They move the witness along from one episode to the next. When a witness has provided evidence on one topic, a transition question lets the arbiter know that questioning on that topic has finished and the evidence on the next topic is to begin, e.g., 'Can you tell us what happened when you arrived at the hospital?'

8.8.3 Order of testimony

Decisions about which witnesses to call and the order in which they are called are dictated by your plan and by the need to develop your narrative. Your witnesses are telling your story.

8.8.3.1 Selecting an order for witnesses

Three principles should be remembered when selecting witnesses:

(a) Call one witness and not more than one corroborative witness on any one point.

(b) Only call those witnesses necessary to prove the elements of your cause of action and to provide the necessary background.

(c) Select those witnesses that are likely to be clear, accurate and reliable.

You should present witnesses in an order that will logically and forcefully present your evidence to the trier of fact. The following principles are suggested by Mauet (in *Fundamentals of Trial Techniques*, see **8.10**):

(a) Present your case in chronological order or in some other logical order.

(b) If your client is to give evidence call him or her first. This reinforces credibility, which may suffer if he or she has been in court while other witnesses have presented testimony.

(c) Follow the rules of primacy and recency according to which people remember best what they heard first and last. Start and finish with a strong witness.

(d) Call an adverse witness in the middle of the case. You will not have started on a bad note, and you will be able to limit the damage done.

(e) An important corroborative witness should normally be called immediately after the primary witness to the facts.

(f) Have proper regard for the convenience of witnesses, particularly experts. Do not keep witnesses at court unnecessarily.

The primary purpose of the order of proof is to present your theory of the case in a logically progressive way, so that it is easy for the arbiter to follow and understand your case.

But remember that last minute problems arise often and you will need to adjust your schedule accordingly.

8.8.3.2 The order of evidence

The basic principle, when eliciting evidence, is to locate the witness in relation to the other characters and events, set the scene, then focus on the action. All necessary preliminary descriptions and information should be elicited before reaching the action because action evidence is most effectively and dramatically presented in an uninterrupted manner.

So a witness, the claimant in a case arising out of a car collision, could be asked questions in the following order:

(a) his or her background,

(b) description of collision location,

(c) what occurred just before the collision,

(d) how the collision actually occurred,

(e) what happened immediately after the collision,

(f) emergency room and initial treatment,

(g) continued medical treatment,

(h) present physical limitations and handicaps,

(i) financial losses to date.

Witnesses should be introduced to the court and will need time to settle. Ask series of questions that identify your witnesses, put them at ease and increase their credibility, for example:

(a) Mrs Smith, what is your full name?

(b) Mrs Smith, where do you live?

(c) How long have you lived at that address?

(d) What do you do for a living?

With a few simple questions you will be able to establish that the witness is married, has children, is a long-standing member of the community and holds a respectable position; all this shows that she is a mature, responsible and, therefore, credible person. Now move on to connect the witness with the case. The easiest way to do this is to use a pointed, leading question, which provides a point of reference but this can only be done when facts are not in dispute. Where there is no dispute it is easy to lead:

(a) Did you witness the car accident in Brooklands Avenue on the morning of 10 May 1993.

(b) Are you the landlord of the premises at . . . ?

If the identity of the person is not in issue, you can begin the testimony there:

How long have you known Mrs Brown?

If the identity of the person is in issue, you can begin the testimony with:

On 10 May were you in the Safeway supermarket at the bottom of Eccleshall Road?

Remember, first set the scene, then describe the action.

8.8.4 Questioning

Having decided on the evidence you will present and on the order in which witnesses will appear you must shape the testimony with your questions. They provide the structure, pace and emphasis. In **Chapter 2** above, 'Interviewing and advising', we discussed the range of questioning possibilities. The more open the question the more a respondent has freedom to use his or her own words. The more closed the question, the greater the control exercised by the questioner. Between these extremes there are a range of possibilities; questions to seek clarification, to gain more detail, and to probe for understanding. Select a questioning strategy for each witness. You can either examine a witness using closed questions, which keeps the witness under control, or you can use open questions to elicit evidence through a free narrative, in which the witness provides all or part of the story in his or her own words.

The decision to control the questioning of a witness in this way depends on the witness, and the nature of the evidence. It is the preferred method where the evidence is important, where caution is advisable, or where the witness might be unable to sustain an account of the facts without detailed prompting. In criminal cases, closed questions are favoured (see M. Stone, *Cross-examination in Criminal Trials*, 2nd ed. (London: Butterworths, 1995), **8.10** above). The witness is taken through the evidence by tightly framed questions, in small steps, and in an orderly and deliberate way. This ensures that all material facts are covered and avoids inadmissible, irrelevant, harmful or prejudicial evidence being given in error.

Too much control in questioning may lead to stark, impoverished forms of testimony, unleavened by rich factual detail or personal involvement. It is for this reason that some writers dislike controlled questions. They believe the witness should be encouraged to tell the story in his or her own words. The best way to achieve this is to use open questions which let the witness tell the story and reveal the important evidence personally. More details and greater emphasis can then be sought using more directed questions.

Paul Bergman suggests the following questioning pattern:

(a) closed questions to establish background and set the scene;

(b) open questions which allow the witness to tell part of the story freely. 'Please tell us everything that happened before she pulled out a gun';

(c) closed questions to bring out details and emphasise that portion of the story;

(d) open questions which allow the witness to tell the next part of the story freely;

(e) closed questions to bring out details and emphasise that portion of the story.

The classification and examples of questions are drawn from a typology developed by Bergman.

8.8.5 The open narrative question

An open narrative question calls a witness's attention to a subject and then asks the witness to talk about that subject. The narrative question provides little direction and imposes no structure. It gives the witness freedom and this gives the testimony a sense of reality and enhances its dramatic effect:

And what happened after your arrived at the wall?

Well, this funny-shaped person kept saying things like, 'I'm going to jump. I'm going to jump.' After a few minutes he pitched forward and fell all the way to the ground. By that time the king's men had arrived on their king's horses, but it was too late.

Narrative questions elicit a more natural form of communication. They can be coupled with specific questions to bring out omitted details:

(a) What did the funny-shaped person look like?

(b) How long after the witness's arrival did he pitch forward?

With a responsive witness, narrative questions can be used at the beginning of the testimony, so that the witness can narrate an overview of a story. Narrative questions should not, however, be used with partisan witnesses, or with witnesses who are prone to digress and fill answers with discrepancies. These kinds of witnesses undermine their own credibility.

EXERCISE 8.3 ROLE PLAY FOR ADVOCATE AND WITNESS

Select a nursery rhyme or other story.

Advocate: ask questions that allow the witness to give evidence about that event using a variety of the questions used for examination-in-chief.

8.8.5.1 Open questions

As narrative questions become more limited, they become merely open questions. Open questions leave to the witness the particular words to be used to describe something. They are distinguishable from narrative questions because here, the advocate selects the subject matter for the witness to talk about:

(a) Why did you approach the wall?

(b) What did he say to you?

(c) What was the next thing that happened?

Open questions move the testimony along. They are short, broad, and non-leading; they do not detract from the witness's own account:

(a) What did you see (next)?

(b) What did you hear (next)?

(c) What did you (he, they) do (next)?

(d) Did anything happen?

(e) What happened (next)?

(f) Then what happened?

Using only short open questions can become boring, so you need to vary the form by using specific, explanatory and follow-up questions where appropriate.

8.8.5.2 Closed questions

Closed questions are those which pointedly ask a witness for a particular fact. They do not suggest the desired answer, but they leave the witness no room to use his or her own words:

(a) What colour was the car?

(b) How long did the incident last?

(c) What are the names of the people you were with?

Each question calls for one bit of information. The benefit of closed questions is that they allow the advocate to control the testimony. Closed questions can be used to 'set the scene', and clarify and emphasise testimony. They can also be used to supply transition points between narrative parts of an examination.

8.8.5.3 Leading questions

A leading question suggests, in its phrasing, its own answer:

(a) Was the man five feet ten inches?

(b) Did he take the purse after he said he would rob you?

(c) The light was on red, wasn't it?

These questions demand a 'yes' or 'no' response. They may assume facts that have not yet been established which is permissible when facts are not in dispute. You should use leading questions sparingly, not because they are prohibited but because they undermine the witness. By suggesting the answer you diminish the impact of the witness.

8.8.6 Special considerations

Sometimes a witness will need to refer to a statement, or notebook, to refresh his or her memory. This is particularly the case with police officers who deal with lots of similar cases. Referring to a notebook or a statement is permitted provided the record was made at the time of the incident or shortly thereafter. You need to establish this and then confirm that the notes can be used:

Q. Can you recall what happened after the car swerved?

A. No. It was too long ago.

Q. Do you recall making a statement to the police shortly after the accident?

A. Yes

Q. How soon after the accident did you make it?

A. About an hour, while I was at the police station.

Q. Was the accident still fresh in your mind at that time?

A. Yes.

Q. Your honour, may the witness refresh her memory by referring to her statement?

EXERCISE 8.4 ROLE PLAY FOR ADVOCATE AND WITNESS

Select a nursery rhyme or other story.

Advocate: ask the witness questions that allow her to give evidence about that event using only leading questions.

8.8.6.1 Real evidence

Real evidence, evidence which is in a tangible form, for example, a weapon, a piece of clothing, a photograph, or a document, provides interest and can have unshakeable credibility. It provides, quite literally, something real and memorable. Ensure that the evidence is admissible (see ss. 1–9, Civil Evidence Act 1968, ss. 23 to 24, Criminal Justice Act 1986 and CPR, r. 33.6(1)). It should usually be disclosed to your opponent and facilities for inspection offered. You also need to establish continuity of possession so as to link each item to the incident.

Each item needs to be marked to identify the person introducing it and enable the exhibit to be distinguished during the trial. The first item will be P1 or D1 and so on, depending upon whether it has been introduced by the claimant/prosecution or the defendant and the judge should be asked to confirm that the exhibit be marked. Finally, the item should be introduced into evidence.

Where there is no dispute, you may lead the exhibit:

A. I saw the defendant with a bloody knife in his hand.

Q. Please look at this knife. Is this the knife?

A. Yes.

Q. Your Honour, may that be marked P1?

Where the item is in dispute it may be necessary to lay a proper foundation for its admissibility or to establish continuity:

A. I saw the defendant with a bloody knife in his hand.

Q. Please look at this (*handed to the witness*). Have you ever seen it before? (*or*) Could you describe the knife?

Q. Is this that knife?

8.8.7 Re-examination

Following their evidence-in-chief, witnesses are cross examined by the other side. When the cross-examination of a witness is complete, an advocate may re-examine his or her witness. The purpose of the re-examination is to give the witness an opportunity to explain, or further develop, matters that were raised during the cross-examination. This means that the scope of the re-examination is limited to what the cross-examiner chooses to raise during his or her examination. It is not another opportunity to go through the evidence provided.

The most common re-examinations arise where the cross-examination has:

(a) called into question the witness's conduct;

(b) brought out only the parts of a conversation or event favourable to the opponent;

(c) highlighted inconsistencies in a witness's testimony or impeached the witness with a prior inconsistent statement; or

(d) shown the witness's testimony to be muddled and confused.

The following two examples illustrate the correct use of re-examination.

(a) The cross-examination of a rape victim has stressed the fact that the victim waited two hours after the rape before she called the police. The implication is that no rape occurred. On re-examination the following question is appropriate:

Q. Why didn't you call the police for two hours?

A. I was upset and afraid. He said he'd come back and kill me if I called the police.

(b) The cross-examination has elicited part of a conversation, that part which helps the cross-examiner. On re-examination the following question is proper:

Q. Other than 'I'm sorry this whole thing happened?' did you say anything else to Mr Smith at that time?

A. Yes, I also said, 'However, if you don't pay me the money you owe, I'll have to hire a lawyer to collect it'.

8.8.8 Cross-examination

Cross-examination is the process of examining your opponent's witness. It is cross-examination which is said to reveal the great advocate, the artist, who is able to win the case through a series of inspired questions. The purpose of cross-examination is, however, often more mundane.

8.8.9 Purpose

The purpose of cross-examination is basically twofold:

(a) *To elicit favourable evidence and advance your own case*—this involves securing the witness' agreement to those propositions that are consistent with your theory of the case.

(b) *To construct a destructive examination which undermines your opponent's case*—this involves asking the kinds of question which will discredit the witness or the witness's evidence.

If you have to cross-examine a credible and truthful witness, you should conduct a constructive cross-examination. You will elicit favourable or indisputable facts. You may seek to develop new emphases and new meanings or seek agreements of alternative interpretations of facts. If you must conduct a destructive cross-examination, you will either challenge the evidence as inconsistent, improbable or unrealistic, or you will challenge the witness as mistaken or untruthful.

Before embarking on cross-examination, you should consider whether it is right to do so. In general, you should not cross-examine a witness whose testimony has not been harmful. If you decide to conduct a constructive cross-examination the opening should reassure the witness, showing that he or she is to be trusted. If you have managed to extract favourable testimony you may decide not to undermine the witness. If you decide you must undermine a witness, elicit favourable evidence first, before you have discredited the witness.

A classic limitation is where a claim or a defence consists of more than one element. If it is conceded that two of the elements exist, but that three other elements do not, consider how far cross-examination is necessary. In a case of theft, for example, you may be prepared to concede that a theft took place but you contend that your client has an alibi. It would be inappropriate to cross-examine a witness whose testimony simply established that a theft had taken place.

In summary, your tactics in cross-examination are dictated by your theory of the case, your objectives or your 'story'. You will elicit a narrative through your own witnesses, but you should seek confirmation of and support for your story by constructive questioning, and only use destructive questioning to challenge competing versions of the story. Your cross-examination should always be informed by a clear sense of purpose.

8.8.10 Elements of cross-examination

A successful cross-examination has a structure that gives the examination a logical and persuasive order. Your chances of conducting a successful cross-examination are maximised when you follow certain rules. The following two subsections of this chapter focus on the structure and the rules, while the third considers the key element in all cross-examinations, the leading question.

8.8.10.1 Structure

To establish a cross-examination that has a clear, logical and persuasive order you need to follow four basic guidelines:

(a) Cross-examination should be restricted to three or four basic points, each of which support your theory of the case. Attempting much more will diminish the impact of your strongest points. Less significant points are likely to be forgotten in any event.

(b) Make your strongest points at the beginning and end of your cross-examination, because first and last impressions are likely to be the strongest.

(c) Vary the order of your subject-matter. The best cross-examinations follow a structure in which your theory of the case is kept implicit rather than explicit. By varying the order of your subject-matter you should be able to achieve your purpose without the witness becoming too aware of the point that you are trying to establish.

(d) Do not repeat the evidence-in-chief. Advocates repeat the examination-in-chief in the vain hope that the witness's evidence will somehow fall apart in the retelling. This invariably fails. The exception to this rule arises when you suspect that the witness's evidence is memorized. In such cases, the witness's repetition of key answers may help you to underscore this point.

8.8.10.2 Rules for cross-examination

The following rules are for guidance, they are not prescriptive but they provide a safe approach.

(a) Know the probable answer before you ask the question. Cross-examination is not a time to fish for interesting information: its purposes are to elicit favourable facts or minimise the impact of the evidence-in-chief. Anticipate the answer every time you pose the question. Alternatively, build up to key questions using neutral questions until you have limited the range of possible answers (see further Boon, *Advocacy*, Ch. 5, para. 10, see 8.10 below).

(b) Listen to the witness's answers. Watch for the nuances and gradations in the witness's evidence and for any reluctance and hesitation. Avoid being caught up in your notes, or worrying about the next question, by organising your notes into cross-examination topics. Avoid following a preordained script and respond to the witness.

(c) Do not argue with the witness. This is both improper, unprofessional and it will damage your own credibility. Do not show frustration with answers not to your liking.

(d) Do not let the witness explain. Open questions allow the witness an opportunity to be expansive and slip in damaging material. Do not, for example, ask how or why something happened.

(e) Keep control of the witness. You can do this by asking precisely phrased leading questions that never give the witness an opening that may damage you. You can also do it by demonstrating to the witness that you know the facts.

(f) Do not ask one question too many. One question too many, the question that explicitly drives home the point alerts witnesses to your theory, allows them to counter it. In your closing speech you can rhetorically pose the question yourself.

This rule is usually demonstrated by some examples, usually apocryphal, of the famous last question.

EXAMPLE 1

Mauet cites the case of a man charged with unlawful sexual intercourse with a girl under 16. A witness giving corroborative evidence was a farmer who said he had seen the pair lying in the field. The cross-examination of the farmer went as follows:

Q. When you were a young man, did you ever take a girl for a walk in the evening?

A. Sure, that I did.

Q. Did you ever sit and cuddle her on the grass in a field?

A. Sure, that I did.

Q. And did you ever lean over and kiss her when she was lying on her back?

A. Sure, that I did.

Q. Nothing improper about that, all perfectly natural and proper?

A. Yes.

Q. Anybody in the next field seeing that might easily have thought you were having sexual intercourse with her?

A. Sure, and they'd have been right too.

EXAMPLE 2

Bergman gives the example of a defendant charged with assault. The cross-examination was of an alleged eyewitness to the assault, whose testimony, succinctly stated, was, 'The defendant bit off the victim's nose'.

Q. Where did the fight take place?

A. In the middle of the field.

Q. Well, where were you?

A. I was on the edge of the field.

Q. How far away?

A. About 50 yards.

Q. What were you doing there?

A. Just looking at the trees.

Q. You had your back to the fight didn't you?

A. Yes.

Q. So the first you knew that there was a fight was when you heard the victim scream?

A. Yes.

Q. And it was not until after the victim screamed that you turned around?

A. Yes.

Q. How can you say, then, that the defendant bit off the victim's nose?

A. Because I saw him spit it out.

In both cases, as Bergman suggests, it is impossible to imagine that the last question would not have been asked of the witness during re-examination, or even during the examination-in-chief. The purpose of the rule is to make you wary of doing more than it is necessary to do.

8.8.10.3 Leading questions

In attitude and verbal approach, cross-examination is the opposite of examination-in-chief. With examination-in-chief, you need to develop testimony in an orderly fashion, and to frame your questions to allow a witness to tell a story in his or her own words so that the testimony will have maximum credibility. In cross-examination your emphasis is entirely different. You have isolated the points that you want to make and you want to make these points clearly and directly. You do not want a witness to tell you the story in his or her own words: that has already been done in the examination-in-chief. You want the witness to verify matters that you have put in your own words. The key to doing this is to use leading questions.

On 13 December 1992, you owned a Ford car, didn't you?
You had two drinks in the hour before the collision, right?
You were looking away from the scene of the accident, isn't that correct?

Leading questions are valued because they focus attention on the advocate, not the witness. Facts are being supplied by the advocate and confirmed by the witness. By phrasing your questions narrowly, asking only one specific fact in each question, you should be able to get 'yes', or 'no', or short answers to each question. This gives you control and leads you gradually to the point you want to establish.

8.8.11 Advancing your own case: eliciting favourable testimony

One of the key purposes of cross-examination is to advance your own case, both by eliciting facts from the witness that will support your case and by putting your theory of the case to the witness. Where you wish to elicit favourable testimony from the witness you should do this first. If you are pleasant and courteous, the witness should relax and cooperate.

It is important to approach each witness's testimony with an open mind. Usually the witness will give evidence upon a number of points that are either neutral to your position or which are directly helpful. The witness may have identified your client but may agree that the crime was committed on a dark night and that the lighting was poor. A witness to a motor accident may agree that the road surface was wet and that both parties were driving fast.

You should review the evidence of each witness and ask:

(a) Are there areas of consensus between the witness's testimony and that of my client?
(b) Has the witness said anything which can be expanded upon to help the case?
(c) Is there anything in a prior statement, which, though not included in examination-in-chief, may help the case?
(d) Has the witness said anything which runs contrary to common sense?
(e) Will the witness cooperate?

The main constructive techniques progress along a continuum. At one end of the continuum you will accept the evidence-in-chief but seek to give it a new emphasis. At the other end you will seek to challenge it by putting your own case. You may, alternatively, seek a fresh interpretation of the evidence by giving it a new meaning (see Stone, M., *Cross-examination in Criminal Trials*, 2nd ed. (London: Butterworths, 1995) on which the following section draws).

8.8.11.1 Emphasis

In the most constructive forms of cross-examination you find something in the evidence-in-chief which favours your case. You accept the evidence, but you make the witness repeat it for emphasis.

The technique could be illustrated by the cross-examination of a sales assistant who witnessed a robbery. The sales assistant claims to have recognised one robber as a customer, by his build, his hair colour and typical way of walking. In his evidence-in-chief, he says, 'It was all over so quickly'.

> What is your strategy for cross-examining this witness?

You might seize on the brevity of the event for emphasis, eliciting further details to build up a picture of how many things the witness seems to have attended to in a short period of time, leaving little time to study the robber. The emphasis on one element, the brevity of the event, supports the cross-examiner's theory of the case: that the identification is either unreliable or wrong.

EXERCISE 8.5

Select a sequence of leading questions to take you through these issues in the cross-examination of this witness.

8.8.11.2 New meanings

Seeking new meanings in the evidence of a witness begins with friendly and indirect tactics which proceed gradually, step by step. By careful leading questions, the witness is made to accept minor adjustments to the evidence here and there, stressing this and toning down that, so that the balance and pattern are subtly altered in the direction of the new meaning. This is where you need to take particular care in posing final questions. Your new interpretation of the evidence may be accepted by the witness, in which case putting the final question should favour your case. But the new interpretation might be denied, in which case you should not put the final question.

These alternatives to 'destructive cross-examination' are considered in more detail in relation to expert witnesses in Boon (Ch. 6, para. 9).

8.8.11.3 Putting your case to the witness

You should put your version of the case to any witness who knows the facts, giving him or her a chance to accept or deny it. This is a general rule of tactics. In particular, if you have led, or intend to lead, evidence to contradict a witness, you should give that witness a chance to explore the contradiction. This can be quite powerful even if you are met with a series of negative responses.

8.8.12 Challenging the opponent's case

Challenging the opponent's case usually involves discrediting the evidence or discrediting the witness.

8.8.12.1 Discrediting the evidence

It is important, when planning a cross-examination, to remember that most witnesses are not lying. They are genuinely seeking to provide an objective and truthful account of the events as they saw them. Attacking them as untruthful is a tactic to be used with caution. You need to focus, instead, upon the manner in which they saw an event. Most people only see an event from their particular standpoint and fill in the gaps in their observations through inferences, connections and conclusions. Once they have done this, they come to believe that this is what they saw. It is your task in cross-examination to explore the gaps between what was seen and what was inferred, to create doubt about the witness's reliability. You should focus on the witness's perception, memory or ability to communicate. As you are not suggesting that the witness is lying, you should adopt a reassuring tone. You should use suitably phrased leading questions to suggest that you sympathise with the difficulties that may have led to mistaken evidence and allow the witness to save face.

8.8.12.2 Perception

Here you are seeking to show that the witness did not really have the opportunity to see the event in question. It occurred quickly or unexpectedly or the witness was too far away to observe accurately.

8.8.12.3 Memory

A witness in court may have be asked to recall events that happened years ago. It can be particularly difficult where the event was an everyday experience and had no particular distinguishing features.

Take a contractual dispute where the defendant has denied receiving the claimant's letter accepting an offer made by the defendant. The claimant's secretary gives evidence that she mailed a letter to the defendant. You may choose to show that, because the secretary types and processes so many letters, she cannot possibly remember this particular letter.

8.8.12.4 The witness's ability to communicate

You may challenge the witness's ability to describe details and directions, to estimate distances and time, and to demonstrate that he or she cannot accurately recreate a picture of what actually happened. By pinning the witness down to specific estimates of distances, time and speed you can often reveal inconsistency with other evidence.

8.8.12.5 Discrediting the witness

You should always exercise caution when seeking to discredit a witness. Attempts to attack the witness can misfire. Of course, if you can show that the witness is not to be believed it can have a devastating effect. There are three ways of discrediting a witness: discrediting the witness's conduct, exposing inconsistencies in the witness's testimony, and impeaching the witness to show that he or she is not worthy of belief.

8.8.12.6 Discrediting a witness's conduct

In this instance you are seeking to show that there is a contradiction between the evidence given and the conduct of the witness. The witness may have given a perfectly reasonable account but may have acted inconsistently. The actions are then invoked as speaking louder than their testimony.

A typical example arises where the defendant in a motor accident leaves the scene of the collision without calling the police. A defendant charged with rape who suggests consensual intercourse may have stayed off work. The witness has acted inconsistently, has not done what common sense suggests he ought. The gap between behaviour and common sense creates doubt about reliability.

8.8.12.7 Exposing inconsistencies in a witness's testimony

One of the most frequently used methods of challenging a witness at trial is by showing that his or her testimony is inconsistent with a prior statement. It may contain a fact which is inconsistent with prior testimony, it may contradict other facts contained in the same testimony, or it may exaggerate or embellish facts. In each case the technique is the same: repeat, build up and contrast (see Mauet, *Fundamentals of Trial Techniques*, see 8.10 below).

First, ask the witness to repeat the fact which he or she asserted in evidence-in-chief, the one you plan to challenge. Use the witness's actual answer when you cross-examine since the witness is most likely to agree with the actual answer, rather than a paraphrase. Then, build up the importance of the statement. Direct the witness to the date, time, place and circumstance of the prior inconsistent statement, whether oral or written. Finally, read the prior inconsistent statement to the witness and ask the witness to admit making it.

Use the actual words. An example, from a civil motor vehicle accident case, illustrates the technique:

Q. Mr Jones, you say you were about 15 metres from the accident when it happened?

A. Yes.

Q. There's no doubt in your mind about that is there?

A. No.

Q. Weren't you actually *over 30 metres* away.

Q. No.

Q. Mr Jones, you talked to a police officer at the scene a few minutes after the accident, didn't you?

A. Yes.

Q. Since you talked to him right after the accident, everything was still fresh in your mind?

A. Yes.

Q. You knew the police officer was investigating the accident, didn't you?

A. Yes.

Q. And you knew it was important to tell the facts as accurately as possible?

A. Yes.

Q. Mr Jones, you told that police officer, right after the accident, that you were over 30 metres away when the accident happened, didn't you?

A. Yes.

Repeat, build up, and contrast. Pick on a simple fact; he was 15 metres away. The contradictory earlier oral statement; he was 30 metres away.

8.8.12.8 Impeaching the witness

A witness may be discredited by showing that his or her evidence was affected by:

(a) bias;

(b) interest in the outcome of the proceedings;

(c) motive for testifying in a particular manner; and

(d) previous convictions or bad acts.

Bias and prejudice are usually established by exposing a family or employment relationship; the witness is incapable of being impartial and objective. A witness may be thought to be partial where he or she has a financial or other interest in the outcome of the proceedings. Greed, love, hate and revenge, are all recognised as compelling emotions. If the witness admits to such an emotion relevant to the case, it is likely to taint the witness, even if the evidence itself is quite plausible. Discredit a witness subtly, bit by bit. If you simply suggest to a mother, who has been asked to appear as an alibi witness, that she is lying to protect her son, she is likely to deny it. Gently build up an impression of partiality:

Q. Mrs Jones, your son was living with you on the date this robbery was committed, wasn't he?

A. Yes.

Q. In fact he's still living with you now, isn't he?

A. Yes.

Q. So you see him every day?

A. Yes.

Q. You talk to him?

A. Yes.

Q. You talk to him about his problems, don't you?

A. Yes.

Q. Wouldn't it be fair to say that you've talked to him about this case?

A. Yes.

Q. You were surprised when he was arrested?

A. Yes.

Q. As far as you are concerned this must be an awful mistake?

A. Yes.

Q. Did your son ask you to come to court?

A. Yes.

Through this gradual lead-up the advocate has managed to suggest the closeness of the relationship, the fact that they had talked about the case, and the fact that any mother is unlikely to believe ill of her son.

8.9 Opening and closing speeches

The importance of the opening and closing address can be overemphasised. Magistrates, judges and juries are unlikely to be swayed by rhetoric. Stick to less ambitious objectives. Your task in the opening is to provide a structure that frames the evidence. In closing, it is to integrate the evidence produced in court with your theory of the case. Before examining the structure and content of these speeches, we deal with some technical considerations.

8.9.1 Who makes opening and closing speeches

In criminal matters the prosecution always have the right to make an opening speech, while the defence usually make an opening speech. In the magistrates' court only one speech is allowed for each party. The prosecution should make their speech at the beginning. In simple cases, and when appearing before a stipendiary magistrate, the right to make a speech is often waived. It may be appropriate in a complex case or where you want to introduce a plan or map in the evidence. In the Crown Court the defence can only open a case if they intend to call a witness to fact, and not in cases where only the accused and witnesses to character are called. The defence opening speech is rare. It should be made after the prosecution have given evidence and before calling witnesses. In criminal matters, however, it is usually only the defence who will make a closing speech. Where the defence exercised the right to make an opening speech then the prosecution will be allowed to make a closing speech. The defence always addresses the court last.

In civil cases it is usually the claimant who begins. In the county court the defendant's advocate may make an opening speech but he or she is not then entitled to make a closing speech except with the leave of the court. As the practice differs between courts, it is wise, if you wish to make an opening speech, to check on the practice in that court. In civil cases the claimant always has the final word.

8.9.2 The opening speech

8.9.2.1 Purpose

The principal purpose of the opening speech is to provide an outline and guide to your case.

8.9.2.2 Opening for the claimant or the prosecution

The opening statement should provide a summary of the evidence that will be produced by the maker of the statement. It should be a summary not an argument. It should be long enough to introduce the case but not too long. You should save details for the actual testimony. Start with a statement of the nature of the case, go on to the issues to be decided, and summarise the facts you seek to establish. The summary of the facts should provide a chronological account. Then describe the harm or loss suffered. You should then move onto a consideration of the pleadings or the counts in the indictment.

If you are opening on behalf of the claimant in a civil case, you could, for example, start by saying: 'May it please your honour, this is a case for damages for breach of contract'. Next, hand the judge the bundle of agreed documents and take him or her through the chronology, tying in the relevant dates with the documents. Build up for the judge a picture of the events in the case. Give a short account of the issues in the case, move on to the statement of case, any schedule of events, and any orders that might have been made. If you wish to amend the statement of case you should do it now. It is best to provide a draft of the amended statement of case and to give appropriate references.

In a criminal case, the prosecution will often forgo an opening speech, especially where it is a simple case. Any speech should be kept brief. Professional arbiters do not need an exhaustive explanation. A jury may need a more detailed account of the evidence that will be introduced. The penultimate step is a summary of the relevant law. The conclusion indicates why, on the basis of the law and facts to be established, you should succeed. In a criminal case before a jury, you will need to explain the law in plain and simple non-technical language.

The opening statement introduces the court to your theory of the case incorporating all the agreed facts with your party's version of the disputed facts, and drawing inferences that you contend derive from the primary facts. An opening need not be boring. It is the ideal opportunity to create interest in the case (see further Boon, p. 76, see **8.10** below). Do not go beyond what the witnesses will actually say. The failure to live up to a promise made during an opening statement may make a case appear far weaker than it actually is. The summary should not be contentious. It should summarise the potential evidence, not the conclusions you want to be drawn or the arguments that you want to make on the basis of the evidence. Conclusions, inferences or judgments of credibility are matters of argument.

The statement should only include a summary of the facts that you can actually produce. If you suspect a witness may not come up to proof, or even not appear, you should not introduce that witness's testimony, nor should you introduce testimony that you hope will be introduced by the other side. Where appropriate you should anticipate the defence. You must work out their theory of the case and try to rebut it. If there are obvious weaknesses in your own theory acknowledge them and explain why, in spite of the weaknesses, the claimant or prosecution is entitled to a verdict. Where possible, you should make your opening lively and interesting. Try to engage the court's interest. Use plain language, try to tell the story as you would to a group of friends. Use eye contact to build rapport. Try to personalise your client, by using his or her name, while referring to the other side as, for example, 'the respondent'.

Napley suggests that the advocate conclude with a summary that provides an agenda; a series of questions which need to be answered in the proceedings. Identify those questions: 'In my submission, the questions you will need to consider when you have heard the evidence are five, and are as follows...'

8.9.2.3 Opening for the defence

As an advocate for the defence you are in a different position to the prosecution or claimant. The trial is underway. Witnesses have appeared and you have disclosed your own position through the questions that you asked in cross-examination. Your purpose is to challenge the case put forward by the other side. But you do not need to destroy their position in opening, you can save arguments about the credibility of the evidence produced, and the inferences to be drawn from it until the closing speech. If you do decide to make an opening you should start with a comment on the evidence given so far. Then provide an outline of the evidence to come, state your theory of the case and anticipate defects in your argument. You should conclude with a summary of the questions that you think need to be answered.

8.9.3 The closing speech

8.9.3.1 Purpose

The closing argument is the chronological and psychological culmination of the trial and your final attempt to address the court. It should integrate the evidence with your theory of the case and present the underpinning argument. It should also rebut or explain the evidence that weakens your case and explain how the law applies. Here you need to exercise caution. If you are addressing a jury, you will need to remember that the judge will follow, and that he may present a different account of the law. You will need to explain your points clearly and in plain terms.

If you have defined your theory of the case and developed clear objectives for the trial, your closing speech should develop logically. If you formulated questions in your opening, now is the time to return to them, to present them again and show how the evidence has provided the answers in your favour. Although you should have planned your speech, you will still need to respond to the evidence given, and to deal with unanticipated arguments. It will be some time before you present a closing speech to a jury, but such a speech is a good model. It provides a basis for your future work and insight into the aims of speeches made by others.

8.9.3.2 Structure

The closing speech should be have a clear, logical structure, be easy to follow and encapsulate the major points. Each case requires different arguments, but most have a similar structure. The following outline provides a simple guide:

Review ↓	Introduction	
	Issues	
	Narrative	
	The Argument	*Analysis* ↓
	Confirmation and Refutation	
	Result	

This diagram suggests two stages. First, you review the evidence using a narrative framework, focusing on the context first, then the action. This is the first opportunity to comment on the reliability and cogency of the evidence. Then focus on your argument, using only those witnesses, exhibits, inferences and techniques that will persuade the arbiter to decide in your favour. Try not to deal with issues of reliability and cogency here unless they are absolutely central to the case. Then, deal with any evidence that supports or refutes the construction you seek to put on the evidence. Do not review the evidence witness by witness, exhibit by exhibit. This can be tedious. Give a structured overview. Organise the evidence around the issues and summarise the evidence that applies to each issue. Situate the evidence in a meaningful context and frame it for the arbiter, to support your standpoint.

8.9.3.3 Introduction

The introduction provides a preview and highlights the most significant and favourable aspects of your evidence. It should not be too detailed, but it should provide a set of prompts. It gives advance notice of your checklist of points. This should help retain interest as you move from point to point.

8.9.3.4 Defining the issues

Use your definition of the issues to control the decision-making process. An implicit definition of the issues will have been provided in the evidence. Now is the time to make it explicit so that the arbiter can easily connect them to the evidence.

Two examples, taken from Paul Bergman's *Trial Advocacy in a Nutshell* (see **8.10**), illustrate the skill of framing an issue so as to produce a perspective on a set of facts. The claimant claims that he was injured as a result of the negligence of the defendant, who claims that he swerved across the road to avoid a small flock of chickens.

The first lawyer, adopting the standpoint of the claimant, frames the issue as follows:

> The issue you have to decide is whether the safety of a bunch of chickens has a higher priority than the life or property of a human being.

How can the defence lawyer frame the issue using the same basic facts?

The second lawyer, adopting the standpoint of the defendant, framed the issue in its rural agricultural context:

> The real issue in this case is whether every chicken farmer in this country has an obligation to erect chicken barricades around his entire property.

The issues which you define may be either legal or factual, or a mixture of both. Often you will define the issues in terms of the legal basis of the claim or defence. In a civil case this will be found in the statement of case, in a criminal case in the indictment. Seek to connect the central question with the particular facts.

In defining the issue you may keep certain issues out, or minimise the damage caused by the negative effect of another issue. Bergman's second example both defines and confines the issue:

> In this case, there is no question but that a particularly brutal crime was committed. Our purpose during this entire trial has never been to claim anything to the contrary. But the issue that is presented to you is the identity of whoever committed such a foul crime. The evidence is overwhelming that it was not my client, Mr Attila.

8.9.3.5 Narrative

It is now time to explain what happened, following the order that you used in conducting your examination-in-chief. Describe the parties, then the scene, then the action: personalise, contextualise, describe. Describe the action from your standpoint, incorporating the inferences that you seek to establish.

Two examples, adapted from Mauet's *Fundamentals of Trial Technique* (see **10.8**), demonstrate the technique. First the prosecution:

> The pub was well-lit. There were lights over the bar, lights over the front entrance, lights in the street that shone through the front windows, and lights from the jukebox. There was more than enough light for the customers to observe accurately and identify the robbers.

Now the defence:

> The pub wasn't well-lit. It was pretty much like any other pub you've ever seen. Successful pubs create moods in large part through dim lighting. This pub was no exception. It had only a few small lights spread throughout the pub. This is hardly the kind of lighting you would want to have when correctly identifying the robbers is critically important.

8.9.3.6 The argument

Here you provide a summary of what happened. You have defined the issues and provided a narrative, now you move on to connect the issue and the narrative:

> Did the police have reasonable grounds for believing that the person driving the car was Tom Jones? The evidence is quite clear. The police had no reasonable grounds

> for believing the driver to be Tom Jones. At the time of the arrest Tom told them he had not been driving the car. You have heard that the police harassed Tom Jones. You have seen that they gave inconsistent evidence. Do you really believe that Tom was driving the car?

8.9.3.7 Confirmation and refutation

Having provided a summary, which again defines the issue, you move on to review the evidence, repeating critical testimony that supports your side and refuting evidence that does not. Selectivity is crucial. Pick the significant parts of the important evidence and argue your case emphatically. When confirming your case introduce your witness, build him up and then review the critical evidence:

> You remember Mr Brown. He was standing at the junction by the Astra. He has been employed at the Post Office for 18 years. He has no interest in the outcome of this case. He does not know Tom Jones. Wasn't he in the best position to see who was driving? Didn't he get a full view of the driver? Wasn't he, in fact, the only person really to see who was driving that car? What did he say? He said...'

The refutation or rebuttal is done in the same way. Select the other side's evidence, minimize it and belittle it. It is now that you challenge the witness's background or qualifications, his or her recollection, or knowledge of the facts.

8.9.3.8 The result

In conclusion you move to a summary and request a proper verdict.

8.9.4 Presentation

Closing arguments should be short, but they should be long enough to cover the ground. To help the arbiters to understand your argument you need to cultivate a slow and clear delivery. Try to maintain eye contact, watch for signs of boredom and respond accordingly. If addressing a judge, see that he or she is keeping up. If the judge is writing notes, slow down; if he or she is content for you to move along, keep to a conversational pace. Keep your head up, not buried in notes, and keep your voice up throughout the sentence.

Use a conversational tone as if telling a story to a group of friends. Generate interest by being enthusiastic. Use themes, raise questions, introduce analogies, slow the topic at appropriate points, make contrasts, provide pointers. Use your theory of the case: 'The real victim is Tom Jones. He is the victim of unreliable identification and the victim of a vindictive campaign.' You then return to this point as you review the evidence. Rhetorical questions can be used to supplement the questions that you have used to define the issues:

> Why should Tom Jones risk conviction? He knows the consequences. As you have heard he has just been promoted at work, he has been married a month, and he has just moved into a new home. Why, at this golden moment in his life, should he risk so much? As you have seen the evidence supports his case.

Analogies and stories can crystallise an idea and connects your ideas with the fact finder's own experience. Analogies do not have to be complex to be effective. Consider these examples cited by David Bergman.

> We have all had the experience of walking down the street and calling to someone we thought was a friend, only to discover we have made a mistake.

We have all seen movies and read books in which innocent people were convicted as a result of mistaken identity.

Remember, present the strongest part of your argument first, then volunteer and deal with your weakness, and finally expose the weaknesses in the position put by the other side.

8.9.5 Afternote

At the conclusion of the case ensure that the issue of costs has been dealt with, including any costs on interim hearings that were left until the final hearing. Always remember to spend some time with the client to explain the decision and its implications.

8.10 FURTHER READING

Appleman, J. A., *Preparation and Trial* (Vienna VA: Coimer, 1967).

Bellow, G., and Moulton, B., *The Lawyering Process* (Mineda NY: Foundation Press, 1978).

Bennett, W. L., and Feldman, M. S., *Reconstructing Reality in the Courtroom* (London: Tavistock, 1981).

Bergman, P. B., *Trial Advocacy in a Nutshell* (St Paul MN: West, 1979).

Boon, A., *Advocacy*, 2nd ed. (London: Cavendish, 1999).

Boon, A., and Levin, J., *The Ethics and Conduct of Lawyers in England and Wales* (Oxford: Hart Publishing, 1999).

Evans, K., *The Golden Rules of Advocacy* (Blackstone Press, 1993).

Hegland, K. F., *Trial and Practice Skills in a Nutshell* (St Paul MN: West, 1978).

Mauet, T., *Fundamentals of Trial Techniques*, New Zealand edition edited by T. Eichelbaum (Oxford: Oxford University Press, 1989).

Murphy, P., and Barnard, D., *Evidence and Advocacy*, 5th ed. (London: Blackstone Press, 1998).

Taylor, N. (Ed.) *The Guide to the Professional Conduct of Solicitors*, 8th ed. (London: Law Society, 1999).

Napley, D., *The Technique of Persuasion*, 4th ed. (London: Sweet & Maxwell, 1991).

Nathanson, S., *Non-Trial Advocacy: a case study approach* (London: Cavendish, 2001).

Stone, M., *Cross-examination in Criminal Trials*, 2nd ed. (London: Butterworths, 1995).

9

Managing your workload

9.1 Learning outcomes

This chapter deals with the need to plan and control your work. Without sound planning we will become ineffective and may well suffer from the effects of stress.

In this chapter we focus on:

- managing your time for personal effectiveness;
- developing a project management approach to legal work;
- key issues of working with and for others;
- managing stress.

9.2 What is meant by 'managing the workload'?

Professional life is very busy, with many demands on people working in professional organisations. Surveys, anecdotal evidence, our own experience all tell us about the pressures and long hours of professional work.

It would be relatively simple if we just had to worry about our 'own' work; that is getting on with the tasks and projects that we have developed or been allocated. But our ability to do that is circumscribed by having to rely on systems and procedures that are not foolproof, on other people within the organisation who may have a different perspective on the work we are doing, and on clients whose objectives may not be clear or may alter over time. Typical examples might be files going missing, computer downtime, inappropriate delegation of work to a junior, clients changing their mind about what they want to achieve, and so on. You can probably add many more of your own examples.

The point really, is that organising to achieve only your own allocated work tasks will inevitably lead you into problems. The unplanned contingencies referred to above have to be taken into account. Clearly we can't plan for exactly when they will occur, but we can take some steps to be ready for them if (or should we say when) they happen.

As a start, try this simple exercise:

EXERCISE 9.1

On what you expect to be a typical day, write down before you start work the tasks you want to work on—a standard 'to do' list. If possible, write down what you would like accomplish with these tasks. At the end of the day check to see how you did. Were you able to achieve everything you set out to achieve? (Congratulations if you did) If not, what stopped you? Were your expectations of what could be achieved unrealistic? Or were there too many unplanned interruptions to your day? If so, what were they?

Some of the likely interruptions you might encounter are:

(a) telephone calls;

(b) trouble shooting;

(c) helping others;

(d) overlong meetings;

(e) colleagues 'dropping in' and outstaying their welcome;

(f) spending too much time dealing with unimportant correspondence and email; or

(g) lack of organisation in others.

Doubtless you can add to these your own pet irritations and time stealers.

Managing the workload, then, is rather more than planning and carrying out the tasks and projects that make up the most important part of our work. It is also about being as ready as we can be to deal with the unplanned, the unexpected and, occasionally, the unpleasant.

9.3 Why is workload management so important?

We have seen that workload management is as much about dealing with the unplanned as the planned. So if we manage to implement some simple initiatives to cope with this mixture it produces benefits for three key stakeholders:

(a) Your organisation. It enables you to add value to your organisation by:
   (i) providing a high-quality service to colleagues and clients;
   (ii) working efficiently, effectively and reliably; and
   (iii) supporting your colleagues.

(b) You and your professional development. It enables you to become more proficient and professional by:
   (i) dealing effectively with a wider variety of work tasks;
   (ii) learning from experience; and
   (iii) avoiding stress and burnout.

(c) Your clients. It will give your clients confidence in:
   (i) your efficiency in dealing quickly and capably with their affairs;
   (ii) your effectiveness in bringing cases to a conclusion as expeditiously as possible; and
   (iii) your reliability in doing what you say you will do within the time and to the standard you have promised.

This calls for approaches in four areas:

(a) managing the time available to make sure that your key tasks are carried out;

(b) managing for the longer term;

(c) working effectively and assertively with others; and

(d) staying healthy and stress free.

9.3.1 How can you manage your available time more effectively?

Time has two of the essential characteristics of any other resource in the commercial and professional world: it is in limited supply and it has a cost. You may find that you are having to stretch the amount of time it takes you to do your work into periods of the week which are normally reserved for non-work activities. You may be spending time at work doing routine, relatively low skilled things that should not be done by someone whose time costs as much as yours. If so, you need to look critically at your time management skills.

Like all organisational skills, time management is about the formation of good habits. It requires a fairly intensive investment at the beginning to understand what needs to be done, but once that initial investment has been made you will find that it is easier to be organised than to be disorganised.

There are four essential steps to managing your time:

(a) find out how you spend your time;

(b) classify your activities into:

- (i) the demands of your job; that is, those things that you must do;
- (ii) the constraints of your job, that is, those features of your working environment which present obstacles to carrying out your job effectively (like having too much to do in too little time, poor supervision, your own lack of organisation, inadequate resources, etc.); and
- (iii) the choices in your work, that is, those areas where you can choose things like when or how or even whether you do something;

(c) map your activities to see how much of your time you are spending on important things, and how much you are spending on things you shouldn't really be doing; and

(d) get organised, so that you can spend your time as effectively as possible.

You have already looked at how your time was spent during a typical day. You now need to do this more systematically.

9.3.1.1 Find out how you spend your time

Try this approach:

(a) Every day for a week, (or for longer if your work has very little routine, but make sure that you choose a representative period) make a list of all the things you want do that day. Indicate on the list when you expect each activity to be completed. For example, if you have to write a letter to a client, take into account any time needed for research, supervision, the postal system, etc.

(b) As you work, make time to log your activities. Remember to log everything you spend your time at work on—so don't forget to note those things which happened that you didn't plan to do.

(c) At the end of your logging period, go through your lists and answer the following questions:

- (i) Which of my planned activities was I able to complete as expected?
- (ii) Which of my planned activities took longer than I expected?
- (iii) Which of my planned activities was I unable to complete at all?
- (iv) What unexpected events occurred which interrupted my planned work?

(v) How accurate was my assessment of what could be achieved in the timescales planned?

When you have completed this activity, you should have a clearer idea of how you spend your time, what obstacles there are to your time management and how long typical activities take to complete.

9.3.1.2 Map your activities

Now that you have good information on what you spend your time doing, you need to find out if you are doing the right things. This is the essential difference between being efficient and being effective. You can carry out your tasks efficiently, but if those tasks are not the most important part of your work, then you are not going to be an effective professional.

Below, you can see the 'time management grid'. This consists of two intersecting continua. The horizontal continuum is the importance line, the vertical continuum is the urgency line. Four quadrants form around the intersection. Try mapping your activities into these quadrants. The importance continuum refers to how much of your time you spend doing the key tasks of your work and how much time is spent doing things that don't really help the progress of your tasks.

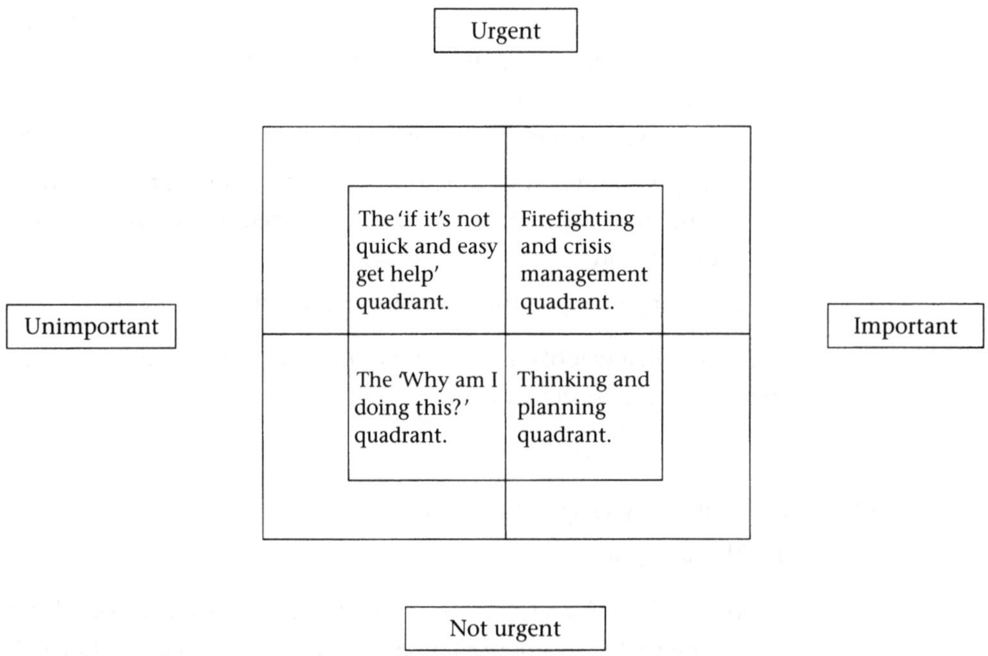

Be careful not to confuse important with urgent. The importance of something doesn't usually change much. Most work tasks become more urgent as time goes by, but that doesn't mean they become more important. Writing a letter to a client or carrying out a negotiation may well be important. If you've left it late to do either of these things, they don't become more important, they just become more urgent. And they certainly become more stressful. You can often judge the importance of something by posing the question, 'What happens if this doesn't get done?' Think of how it affects your client's interests, the reputation of your firm and your own professionalism.

When you have plotted your activities, note the quadrants in which your important tasks are being carried out.

If you have a lot of plots in the 'firefighting' quadrant, it usually means that you don't have enough time to plan your work properly. This can often mean that mistakes are made, important elements of the job are overlooked and you have to spend time later rectifying problems. Although many people get a buzz out of working in the pressured atmosphere of this quadrant, it also leads to high stress levels and increased workload and is unlikely to endear you to clients and colleagues if you get a reputation for leaving tasks until the last minute.

If most of your plots are in the 'planning' quadrant, congratulations. You are clearly able to put in the appropriate level of planning and preparation. Most effective professionals do most of their work in this quadrant, so aim to stay there.

However, if you find yourself working a lot in the 'why am I doing this?' quadrant, you are probably underemployed and probably bored. Why are you spending time on unimportant things? What is being left undone while you are doing these things? The implication is that if you spend a lot of time doing things which are not central to your work, then this limits your ability to plan effectively to do those things that are. Consequently, when you come to do the important things, you are probably doing them in the 'firefighting' quadrant (see above).

Finally, if you are pressed for time when doing unimportant things, you probably put too many of your plots into the 'urgent but unimportant' quadrant and you could use some help. First of all, what proportion of your work activity is spent in this quadrant? Secondly, why are you working this way? This quadrant tends to be stressful because everything is urgent and mistakes can be made. It also tends to be rather unsatisfying, because you are not achieving anything important.

It is most likely that you have a scatter of plots in different quadrants at different times. Try to see if there is a pattern. For example, is there a tendency for you to work in the 'why am I doing this?' quadrant when tasks you don't like crop up? Or are you, as a subordinate, having someone else's trivial tasks delegated to you? Is your lack of organisation making you less effective than you would like to be?

If so, the next section will give you some hints on how to be better organised and how to deal with some of the things which typically steal our time.

9.3.1.3 Get organised

Getting organised falls into four main areas:

(a) Use a diary.

(b) Use a personal planner.

(c) Inform other people of your time management processes.

(d) Learn to manage the 'time stealers'.

Use a diary

This is probably the most obvious step to take for anyone wanting to improve their time management. All professionals keep diaries. Many practising solicitors need a system for recording time spent on activities relating to specific cases and it may be that you can incorporate your personal time management around this system. So much the better if you can. If you can't, because, for example, it is software based and you are unable to customise it adequately for your own use, then you need to develop your own diary system.

At its simplest any diary system needs to do the following:

(a) Record your appointments. Ideally, your appointment entries should record who the appointment is with, when and where it will take place, what is to be discussed and how much time has been allowed for it. It is useful to agree this last item when you first make the appointment. The people you are meeting can then plan their time and you don't risk compromising other tasks or other appointments because meetings or interviews have overrun.

(b) Show tasks carried forward. Obviously not all the tasks on your 'to do' list will be completed in one session. It therefore makes sense to show how and when they will be carried forward and record this in your diary. Take care when you do this to allow for any preparation time needed before you follow up the task. For example, if you have written to arrange a meeting with a client, you may put a reminder in your diary to check if the client has responded to your letter after, say, a week. Or, if the client has responded, you may want to remind yourself to read the file or carry out any necessary research in plenty of time for the meeting.

(c) List critical dates and times. The nature of legal work demands that certain tasks are carried out within certain critical timescales. Make sure that these critical points are recorded in your diary, again allowing for any preparation time.

Use a personal planner

These come in many forms. There are many computer-based applications, which can combine the functions of planner and diary, like Microsoft Outlook or Lotus Organizer. Hand-held electronic organisers, which carry out the same functions, are becoming more and more popular.

> What do you think are the benefits and disadvantages of using computer applications to plan and organise your work?

If you use electronic planners, you are at the mercy of your power supply or the reliability of your organisation's network. You may also need to get into the habit of putting all your appointments onto your computer or transferring information between your hand held organiser and your desktop.

Using electronic means to plan and organise your work has many benefits, but there is the initial effort of learning the system and making sure you develop good habits of logging your information and backing it up.

Many organisations now use computer-based time logging software to track fee-earners' work and as the basis for billing clients. If this is the case, then you need to comply with the system in use.

If you prefer to stick to pen and paper for this task, there is a wide selection of items you can get at high street stationers. Some people like to see long periods at a glance, for example, so they use wall charts. These are very useful ways of identifying busy and slack periods and avoiding 'double-booking' your time. Wall charts are especially useful for noting major projects (see **10.3.2**), closure periods and holidays.

Planning falls into immediate, medium-term and long-term plans. Immediate plans can be dealt with by making a 'to do' list every day. Don't forget to estimate the time you need to spend on these daily tasks and to prioritise them. You should also get into the habit of reviewing your 'to do' list at the end of each day to see how well you are managing your priority tasks.

Medium-term plans refer to the forthcoming few weeks or maybe the next few months. These plans deal with the progress of ongoing projects (see **10.3.2**).

Long-term plans refer to the next year, or whatever convenient or customary period your organisation uses, or the project requires. For example, if you are a student your long-term period may be the academic year, the semester or the term.

Planning long-term may well be more speculative than immediate or medium-term plans. However, you can still put in things which are fixed, and estimate other likely events.

Inform other people of your time management processes
Planning your time is fine. Sticking to your plans is sometimes more difficult. The aim, remember, is to do your important work in the 'thinking and planning' quadrant. This is crucial if your important tasks are to be thought through carefully and carried out effectively. This means that you have to make the time to think and plan. Which is, of course, difficult to do in a busy workplace. It is useful, therefore, to accustom your colleagues to the fact that there are periods in the working day when you need to work uninterruptedly. You may be in a position to identify a period each day when you can do this routinely. It is more likely, however, that your work schedule prevents this. Nevertheless, you should be able to recognise from your planning system when you need time to work without interruption. The key here is to communicate this to your colleagues. It is quite possible to have all phone calls re-routed, or to use an answering machine or have someone list your calls. You can then return the calls when you have finished your high priority work. Your colleagues should accept this approach, since they too will need similar periods of time to think and plan.

> You have decided that you are not getting to grips with your increasing workload. Therefore you decide to spend thirty minutes each day thinking about and planning high priority work. You can't use the same time each day because the nature of incoming work is unpredictable, so each week you pencil in a likely period each day and let your colleagues at all levels know. Occasionally, this has had to be altered because it presents problems for others, but you have always been able to renegotiate the necessary time. Your colleagues respect your wishes and leave you in peace to work during that time. Only the most obviously urgent calls are put through from the switchboard.
>
> Your senior, however, is constantly interrupting this period with enquiries about your progress on various tasks or with additional, and often urgent, minor tasks. These tasks have often arisen because of her own inability to manage her time effectively.
>
> What can you do about this?

This kind of situation calls for a certain amount of tact and assertiveness. You need tact because this person probably has every right to delegate such work to you. You need to be assertive to point out that if you stop and do the work they have given you, then you will have to postpone doing the more important work. This may not cut much ice, of course, because your senior may well confuse urgent with important work and tell you to do the urgent work first.

A way a dealing with this is to speak to your senior at a convenient point in the day. Mention the time you have planned for your 'quiet' session and ask outright if there are likely to be any other small tasks to do during the day, so that you can get them out of the way, or plan an alternative time to do them.

Learn to manage the 'time-stealers'
It would be impossible to engineer your working day so that there were no interruptions or unplanned events requiring your attention. The issue, therefore, is to look at how you deal with such things.

Look again at your time log. What things cropped up to obstruct your work plan? Which of them arose because of what others did? Which of them arose because of your failure to perform effectively? How did you deal with them? You could ask the same question of any professional and you would find the same kind of things on their list.

Below is a table of some of the main time-stealers found in professional life, with some suggestions on how to manage them.

Problem	Possible solution
Unclear responsibilities	Discuss with your senior(s) at regular times what they regard as your work priorities. If they are not organised enough to tell you in detail, make a plan yourself and use it to negotiate with them.
Personal disorganisation	Establish priorities, plan your work, manage your time. Re-train yourself into new, more effective routines. Set yourself achievable goals.
Not enough time to do the research needed to perform effectively	Establish clear priorities. For important matters, like research, set aside time to do it and inform others what you are doing.Examine and reflect on your research skills.
Too many interruptions from telephone calls	Separate important from unimportant. For important calls decide if they can be dealt with there and then. If not, call back when you have had time to give the matter appropriate consideration. For unimportant calls, you may need to have a receptionist filter them. They can then be listed for you to call back later.
Unrealistic deadlines	Everything takes longer to do than you think. Aim to leave about 20 per cent of your time available to deal with unplanned events. It is not possible to plan every moment of your day.
Postponing unpleasant or difficult tasks	Plan to do them first and get them out of the way. You will feel better.
Fear of making mistakes	Whenever you are unsure, ask. Put forward your solution and ask for feedback and constructive criticism.
Poorly led meetings	Make sure you are clear about what has been decided, and why. If necessary, keep your own unofficial minutes.
Waiting for important information	Depending on how important it is, decide how necessary it is for you to make a nuisance of yourself to get it. Learn to give other people deadlines for giving you information.
Interruptions by visitors	Communicate times when you are not available. At other times make a judgement between maintaining good relations with colleagues and having your work held up. Learn how to terminate informal visits politely and unambiguously.
Can't say no	Say no. Go on, try it. Avoid becoming the dumping ground for everyone else's problems. Learn not only to say no, but to say it clearly, without being defensive and giving lots of weak excuses.

This list is only a guide. What can be achieved will depend to a large extent on the climate within your organisation. You may work in a very authoritarian environment, or it may be very disorganised. Each organisation is unique in accepting certain behaviours and discouraging others. In your organisation, some of the suggested solutions may be unacceptable. However, to be effective and professional, you need to manage your time efficiently. Look at the list and see whether the suggested way of managing the problem is feasible in the context of your organisation. If not, what else can be done? Where necessary, add to the list from your own experience and try to put in ways of managing that problem.

9.3.2 Managing for the longer term

Many professionals who are involved in case work take a 'project management' approach to planning and managing in the longer term. The time management approach taken above is fine for the day-to-day management of your workload. But cases which may take years to come to a conclusion require a different approach. This is particularly true if you have a big caseload, where it is difficult to recall in detail the precise stage at which any case may be.

Project management can be as simple or as complicated as you want to make it. However, any approach you take needs to do four things:

(a) Identify all the contributory activities that make up the project.

(b) Set appropriate time periods for each activity.

(c) Show whether activities are dependent on, or independent of, other activities.

(d) Link to your time management system.

A straightforward approach is to base your project planning on a Gantt chart. This is a simple, basic document that you can customise to your own needs very easily. Although simple, it nevertheless satisfies all the criteria listed above.

To create a Gantt chart for a case, or project:

(a) Decide on the most convenient time scale. You may decide on a yearly plan, with monthly units. What you decide depends on the state of your knowledge of the project and the critical points that you know will arise.

(b) Identify all the contributory activities which are needed to complete the project.

(c) Across the top, enter your timescale.

(d) Down the side, enter each of the activities.

(e) For each activity enter the point at which it begins and the point at which it ends. Join the two points with a line.

(f) Where an activity depends on the progress or completion of other activities, link them with vertical lines.

(g) For each month, or other convenient period of your timescale, transfer any tasks that contribute to the completion of an activity to your diary or to-do list.

Look at the illustration below:

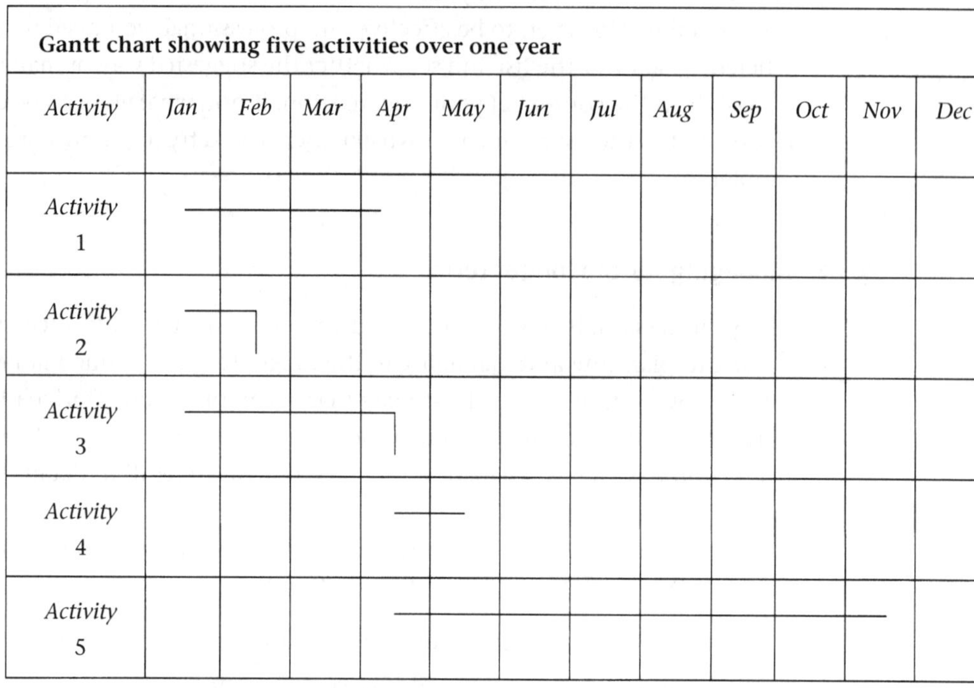

What this chart shows are five activities. The first three begin at the same time. Activity one is independent of the others. The result of activity two is needed to complete activity 3. For example, activity two might be arranging a medical report on a client. The results of that report will feed into activity 3, which may be periodic contact with your client to gather evidence for a personal injury claim. Once all the appropriate evidence is gathered, you will issue a Claim Form to initiate proceedings.

Activity 5 is another activity independent of the others. Naturally, this is a very simplified example. A legal case will inevitably be rather more complex than this. However, its increased complexity is because of the greater number of activities than those indicated here. It is not because it is conceptually more difficult. Using these simple principles, you can plan and track any project. You can also make your chart much more sophisticated by colour-coding the activities according to what they are. For instance, you might wish to code activities involving the other side with its own colour.

Any of the activities is made up of a number of tasks. These might include writing letters, holding meetings and interviews, attending hearings etc. These tasks should be integrated into your day-to-day time management system, leaving enough time to complete tasks in order to meet deadlines on your project plan. Remember the '20 per cent rule' alluded to at **9.3.1.3** above. Leave yourself at least 20 per cent of planned time for delays and other unforeseen eventualities.

If you haven't yet succumbed to planning fatigue, there is one more document you might wish to incorporate into your system. This is just a checklist of all the activities in your Gantt chart, with the completion deadline shown. Tick off each activity as it is completed. This ensures that all activities are completed to plan. This last is a very simple, but useful, document which gives you a sense of security because you see your project being completed within the planned time.

9.3.3 Working effectively with others

A feature of modern professional life is the need to work effectively with colleagues, other professionals and members of other occupations.

> What do you think are the benefits of working with others? List them and compare your list with that of a colleague.

There are several perceived benefits of working with other people:

(a) to share experience and expertise;

(b) to work more effectively by sharing out work tasks; and

(c) to give and receive emotional support.

9.3.3.1 Working with colleagues and seniors

You will usually find yourself working on a day-to-day basis with at least one other person (usually your senior). Effective working of this kind calls for as much clarity as possible on the part of each as to the expectations they have of the other. In a professional organisation there is no room for 'mind-reading'. A senior cannot expect a subordinate to second guess what is in his mind. However, after working effectively together for some time, expectations become norms of behaviour and do not have to be constantly restated. When expectations do alter, however, that is something which needs to be made clear and discussed.

> When work is delegated to you, what do you need to understand about the task?

It is most likely that you will want to know:

(a) what the task is;

(b) what outcomes are being sought;

(c) what resources are allocated to the task;

(d) the timescales;

(e) who else is involved; and

(f) who is accountable.

If this doesn't happen, you may need to take the initiative and ask the relevant questions. If the answers you get are too vague, then you may need to set your own objectives for discussion. Do this by outlining 'SMART' objectives. This is an acronym to help you set suitable and effective objectives for any task. It stands for objectives which are:

(a) *specific*—and leave no room for ambiguity;

(b) *measurable*—so that you know when your task is satisfactorily completed;

(c) *attainable*—so that the goal is achievable with resources available (including the resource of your capability);

(d) *relevant*—so that the objective contributes to the achievement of overall goals in a transparent way; and

(e) *timed*—so that you know the amount of time available to complete the task.

Problems sometimes arise with colleagues or seniors when there is confusion or disagreement about roles. A role is set of expectations that someone has of a person operating in a particular relationship to them. We all have different roles in our lives. There are expectations of us as brothers and sisters, parents, bosses, subordinates, etc. The climate of the organisation or workgroup influences the set of expectations in working relationships. Unless these are made clear, then there is scope for role conflict or role confusion.

How can you avoid this, or do something about it once it happens?

EXERCISE 9.2 RELATIONSHIP ANALYSIS

Identify the person with whom you want to improve relations.

(a) Write down what your main expectations of that person are.
(b) Then write down what you believe their expectations of you should be.
(c) Fix a time to discuss this with that person.
(d) Listen to what they have to say.
(e) Adjust your list of expectations in the light of your discussion.
(f) Both parties set objectives to change their behaviour.
(g) Review after a suitable period.

Such an initiative may be totally unreasonable in certain organisational climates. But where it has been done, it has delivered several benefits:

(a) It clarifies expectations on both sides.
(b) It opens channels for dialogue about roles and performance.
(c) It requires both parties to reflect on their relationship with each other and with others.
(d) It tends towards creating a climate of trust and openness.
(e) It enables the giving and receiving of feedback in a positive spirit.

9.3.3.2 Working in groups

Depending once again on the climate of the organisation you are working in, you may be required to work collaboratively with others. These may be colleagues, or other professionals on cases needing multi-agency involvement. In the former situation, the management of the firm will be looking to achieve certain benefits from this approach.

Effective groups produce synergy. That is, the output of the group is greater than the sum of the individual contributions. It is therefore no wonder that organisations find the notion of group working attractive. However, in order for groups to deliver synergy, there needs to be investment of effort in the development and maintenance of a work group. This can be quite costly of people's time. Nevertheless, it is a necessary investment since, without attention to these things, you may end up with an ineffective group, in which the group's output is actually less than the sum of the individual members' contributions.

EXERCISE 9.3

Think of one effective group you have belonged to and one ineffective group you have belonged to. These need not necessarily be work groups, though it is better if they are.

Try to identify and list the features of each that made it effective or ineffective. Compare your list with that of a colleague.

Effective groups have most of the following characteristics:

(a) shared commitment to the goals of the group;

(b) appropriate attention to the goals to be achieved, the emotional needs of the group as a whole and the emotional needs of each individual;

(c) accepted standards of behaviour among group members for things such as decision-making, resolution of conflict, giving and receiving feedback;

(d) a dynamic and effective mix of roles within the group; and

(e) a climate of openness and trust among group members.

These characteristics do not develop haphazardly. There are identifiable stages in the development of effective groups and identifiable processes in keeping groups working effectively.

Tuckman ('Development Sequence in Small Groups', 1965) identified the four stages of group formation:

(a) *forming*—where the group comes together as a unit with collective goals and tasks;

(b) *storming*—where ideas and personalities are challenged and individuals jockey for positions within the group;

(c) *norming*—where the group evolves norms (or rules) of behaviour. These are usually about how decisions are made, conflicts settled, etc.; and

(d) *performing*—where the group carries out its tasks effectively and collaboratively.

The addition of a new member can upset a performing workgroup, so that it has to go back to one of the earlier stages of development, normally the 'storming' or 'norming' stage.

Effective groups reach the performing stage quickly. This is because there is the most appropriate mix of group roles in that collection of people. Belbin ('Team Roles at Work', 1997—see **9.3**), developing his earlier work, identified nine roles needed to make an effective group. Not all of these roles are needed all of the time. Nevertheless, any effective group needs people who can take on the required roles at the appropriate time. Most of us have the ability to take on several of these roles. We are more comfortable with some than with others, though, and tend to opt for the ones we are comfortable with.

> What roles are necessary, in your view, to make an effective group?

Belbin has researched this matter for more than 20 years. During that time he has gradually refined the number and descriptions of team roles. His most recent list of roles (1993) is:

(a) *Plant*—this is the creative, often unorthodox thinker in the group, the one person who can be relied on to come up with a novel suggestion.

(b) *Resource investigator*—this is the enthusiastic extrovert, who is a skilled networker. This group member often has many links outside of the group.

(c) *Co-ordinator*—this is the member who exercises control over all the activities of the group; it is the leadership role.

(d) *Shaper*—this is the member who drives the group forward, focusing on the achievement of the group's goals.

(e) *Monitor-evaluator*—this is the member who applies an analytical mind to the group's task. Monitor-evaluators challenge ideas and force others to think things through carefully.

(f) *Team-worker*—this is the member who pays most attention to the emotional balance of the team; they tend to be cooperative, diplomatic and calming.

(g) *Implementer*—this is the member who is always looking for the best way to put ideas into practice.

(h) *Completer*—this member is the details person who makes sure that objectives are properly completed and on time.

(i) *Specialist*—this is the member who provides the required expertise in the group.

Clearly not all of these roles are needed all the time. Nor are we to infer that an effective group should have nine members. We all have our preferred roles, and provided that all the roles needed are available, then the number in the group may vary.

> Try to think of contexts in which solicitors need to work together as a group.
> How might each of the Belbin roles be realised in those contexts?
> What are your own preferred roles?

The Tuckman group development process and the Belbin roles provide us with useful analytical tools for identifying problems in group working. Presenting a framework like this gives group members a shared language in which they can discuss problems of group dynamics and ineffectiveness. The Belbin questionnaire is widely available and is a useful tool to provoke discussion on group issues.

9.3.4 Coping with the pressures of professional life

Evidence suggests that stress at work is on the increase (John Arnold et al., *Work Psychology: Understanding Human Behaviour in the Workplace*, 1998—see **9.5**). By stress we usually mean a level of pressure that we find difficult to cope with. Professional workers are just as susceptible to stress as any other worker and it is important that we recognise what stress is and learn to do something about it. There are two key points to bear in mind:

(a) Stress is very individual—what one person copes with without difficulty, another may find extremely stressful.

(b) Continually being in a situation you find stressful will have serious consequences for your emotional and physical health.

9.3.4.1 What stress is

Cummings and Cooper (1979—see **9.3**) developed a framework demonstrating the process by which stress affects the individual. They suggest that individuals try to keep their thoughts, emotions and relationships with the outside world in a 'steady state'. Each element of a person's physical and emotional make-up has a 'range of stability'. While the individual remains within that range, they feel comfortable and able to cope. When some force or other pushes one of these elements outside the range of stability, the individual needs to take action to return to the stability range. The behaviours associated with returning to the range of stability are referred to as our 'coping strategies'.

Therefore, any pressure that pushes us beyond the stability range causes us stress. Clearly, if we cope adequately with such stress, then it is not a problem. If we don't cope, it is a problem, and a potentially serious one.

> Think of a time when you have felt unable to cope. What brought on your feelings? What did you do about it?

9.3.4.2 Causes of stress

Arnold et al. (1988—see **9.3**) identified clusters of factors which contributed to workplace stress:

(a) *factors intrinsic to the job*—workload, long working hours, the physical working environment, the effect of your decisions on others, etc.;

(b) *your role in the organisation*—role ambiguity, role conflict, susceptibility to anxiety, etc.;

(c) *relationships at work*—lack of support from seniors and colleagues, unfriendly atmosphere, too great a level of competitiveness, etc.;

(d) *career development*—progress not according to expectations, job insecurity, feelings of inequity in treatment; and

(e) *organisational structure and climate*—degree of influence in organisation, incompatible climate in organisation, etc.

Remember that these factors trigger stress in some people and not in others. So we need to bear in mind that how the individual perceives these factors and how they cope with any they find stressful, will ultimately determine how much of a problem they are.

Moreover, once the stability range has been exceeded, other factors, which the individual coped with previously, may well become sources of stress themselves. This then makes it difficult to isolate the major stressor. Approaches to managing stress need ideally to deal with the organisational causes of stress as well as with the individual suffering from stress.

9.3.4.3 Managing stress

Arnold recommends a five-step approach to managing stress:

(a) Be aware and accept that a problem exists—being stressed is not an indication of incompetence.

(b) Be able to identify/isolate the problem/stressor—Arnold et al. suggest that when you become aware of a problem, you keep a diary of what happens to make you stressed, and how you attempt to cope with it.

(c) Attempt to change the problem/stressor in a way that provides a solution which is mutually beneficial—this may be possible if it involves issues of role conflict, or workload management, as described elsewhere in this chapter.

(d) If the problem/stressor cannot be changed, then find a way of coping with the problem—the things that stress you may not be work related, or stressors outside of the workplace leave you less able to cope at work. Backing from colleagues and friends on an informal basis is probably the most common form of support. However, there are numerous counselling services available for more professional support.

(e) Monitor and review the outcome—evaluate your responses to stress, reflect on them and plan to deal with other stressors in similarly effective ways.

9.3.4.4 Staying healthy

Stress has unhealthy consequences because stressors trigger defensive physical mechanisms which were naturally selected to help human beings in physical danger to respond to those dangers. The triggering of those mechanisms requires a physical response (fighting or fleeing) which uses up the adrenalin secreted at the moment of stress.

Since professional life takes place in the confined space of the office, fighting or fleeing is not easy! Moreover, hitting something or someone or running away probably won't make things better. If we experience these responses frequently, and fail to find a physical response, it is not surprising that over time, stress can have serious physical consequences.

9.3.4.5 Take exercise

One way of coping with this is to take regular physical exercise. The consequences of this are threefold:

(a) You dissipate toxins brought on by stress.

(b) You become generally fitter and healthier.

(c) Because you are fitter and healthier, your body is able to deal more effectively with stress responses.

9.3.4.6 Identify your stability zones

You can also identify the 'stability zones' in your life. These are those areas of your life that are the most stable; these are often our family relationships and friendships. They provide us with an environment where we can talk about problems and know that we will receive a sympathetic hearing. They provide emotional support and practical help.

9.4 Conclusions

Managing your workload involves five things:

(a) organising yourself by identifying how you use your time and reorganising your working practices to make the most of this scarce, expensive resource;

(b) where possible, seeing your workload as a series of projects with time-constrained activities. This enables you to keep track of a heavy case load and ensure that no critical deadlines are missed;

(c) avoiding role confusion or conflict by clarifying expectations and accountabilities;

(d) analysing the dynamics of any group you work in to make sure that it performs as effectively as possible;

(e) taking measures to reduce the effects of stress.

This chapter has been an introduction to some of the key issues in managing yourself to perform effectively. Below is listed some further reading that you might find useful. Remember that, although the chapter has focused on improving performance in the workplace, you can just as easily use the material to improve the management of your study workload on the LPC. As a student, you still have to balance study with other aspects of your life. You have a very heavy workload and your time is in limited supply. If you think about it, you might achieve a significant degree of synergy by working collaboratively with fellow students.

Finally, workload management is relatively easy if you pay attention to it at the beginning of your career, when the workload is limited and you don't have to train yourself out of too many bad habits. At that point the investment of effort is smaller and the return greater.

9.5 Learning outcomes

You should now be able to:

- analyse how you spend your time;
- identify the key tasks of your work;
- establish priorities in your work;
- organise your work so that high-priority work is planned and executed on time;
- identify the things that reduce your effectiveness in carrying out your priority tasks;
- determine time-scales for new work;
- take a 'project management' approach to carrying out your work;
- carry out a process to improve working relationships;
- describe the processes by which individuals can collaborate effectively;
- state the requirements of effective delegation;
- take appropriate steps to manage stress.

9.6 Self-test questions

(1) What key documents will help you develop your time management skills and what is each used for?

(2) What are the four key elements of project management and how can a Gantt chart help?

(3) When a task/project is delegated to you, what six things is it important to be clear about?

(4) What factors does the research of Arnold et al. (1998) show as being the main causes of stress?

9.7 FURTHER READING

Arnold, J., Cooper, C. L., and Robertson, I. T., *Work Psychology: Understanding Human Behaviour in the Workplace*, 3rd ed. (London: Financial Times Pitman, 1998).

Belbin, R. M., *Team Roles at Work* (London: Butterworth-Heinemann, 1993).

Belbin, R. M., *Changing the Way We Work* (London: Butterworth-Heinemann, 1997).

Cummings, T., and Cooper, C. L., 'A Cybernetic Framework for the Study of Occupational Stress', *Human Relations* (Vol. 32, 1979, pp. 395–419).

Maitland, I., *Managing Your Time* (London: IPD, 1999).

Tuckman, B. W., 'Development Sequence in Small Groups', *Psychological Bulletin* (Vol. 63, 1965).

10

Continuing your learning

10.1 Introduction

This chapter deals with the processes of continued learning and development throughout your career. One of the features of professional life is the need constantly to update your knowledge and skill and to develop your expertise as a professional. The principal way that you will do this is through your day-to-day work. In other words, you will learn by making something of your experience.

To some extent we learn simply by having experience, but that kind of learning is haphazard and patchy and largely happens without our being aware of it. Unfortunately, this means that we are unable to build on our experience in any focused way. Experts become expert by taking responsibility for and control of their learning. They are thoughtful and reflective about the things they have done and observed, and they make clear plans to improve their performance.

In this chapter we focus particularly on:

(a) how experiential learning happens; and

(b) how you can use your experience to become more competent.

10.2 How experiential learning happens

As a student at law school, you are presented with a syllabus in which knowledge and skill are segmented, and your ability to apply the content of these segments to problems in prescribed ways is what is assessed.

In contrast, professional, experiential learning does not come to you in convenient, ready-made segments. Frequently, there is no clear-cut problem confronting you, with a corresponding, clear-cut answer. Nor is anyone going to give or withhold marks for the quality of your answer. The main criterion by which you will be judged will be the degree to which your clients achieve their objectives.

It also follows that there is no end point to your learning. There is no stage at which someone says to you that you are now a fully-fledged expert.

So you can see that the learning process you need to adopt in practice is significantly different from the one you will have followed previously.

How then do we learn from our experience? Following a long line of thinking in learning theory, David Kolb (*Experiential Learning: Experience is the Source of Learning and Development*, 1984—see **10.4**) suggested that we take a cyclical approach to our experience in order to learn from it. He identified four stages in his cycle.

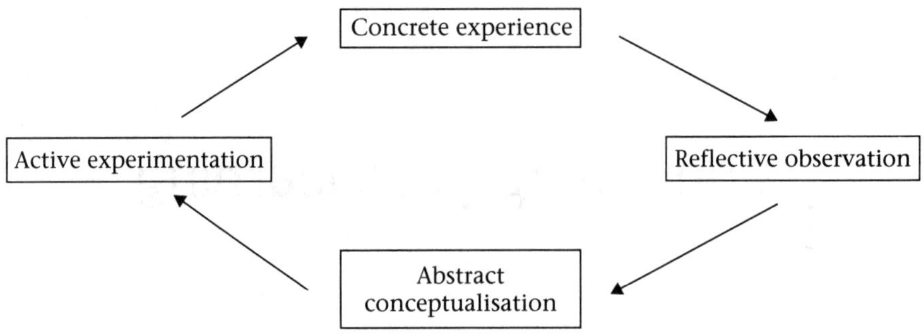

10.2.1 Kolb's Experiential Learning Cycle

The first stage is 'concrete experience'. This is quite simply an event or sequence of events which we take part in and from which we want to learn. 'Reflective observation' is the stage at which we try to make sense of our experience through thinking about what happened, why it happened, why we opted to do one thing rather than another, how satisfied we were with the outcome, etc. We can fruitfully base our reflection on three aspects of our experience:

(a) what we did (our behaviour);

(b) why we did it (our thought processes); and

(c) our feelings and attitudes towards what we did.

Once we have understood these things, we move on to the next stage, 'abstract conceptualisation'. This is the stage at which we integrate our reflections into the wider framework of our understanding of the kind of events we experienced at the start of the cycle. In professional life, our actions are based on countless 'theories of action'. These are beliefs about what behaviour will produce the most beneficial consequences in given situations. The abstract conceptualisation stage is the point at which we create a new theory of action, or, more likely, modify an existing one.

The last stage of the cycle 'active experimentation' implies that we consciously plan to test our theories of action to see if they produce the outcome intended. We then put this theory into practice when a similar situation to the original next arises. The cycle then continues, leading to a gradual, incremental refinement of our theories of action.

Look at the following scenario which illustrates the process:

> You have just interviewed a client and as you are writing up the interview notes, it occurs to you that you forgot to ask for some important information. You therefore need to contact the client, explain the situation and obtain the information.

Reflection on this event may well take the form of mentally replaying the interview, identifying where you missed your opportunity to get the information and recognising your embarrassment at possibly appearing incompetent to your client and to colleagues.

You may conceptualise this in several ways, depending on what is most important to you at the time. You may determine that in future interviews of this kind, you will make sure you get any similar information. You may decide that the error isn't really so bad, because you can get the information at a later meeting, so that you can be a bit more relaxed in future if you don't get all the information you need.

In the experimentation stage, you may well rehearse in your mind how you are going to remember to get the information in future interviews, or that you are going to make a note to remind yourself.

The process, when we write it down, seems much more long-winded than it usually is in practice. For instance, your reflection may slip rapidly and imperceptibly into the conceptualisation stage. Following a mistake like the one in the illustration, it takes a matter of a few moments to modify your theory of action and to think about how to improve matters.

Moreover, as you become more confident, you learn to recognise which events afford the richest opportunities for learning. You begin to set your own learning agenda and when that happens, you are beginning to take control of your professional development.

How effectively you learn from experience depends not only on selecting rich experiences, however. It also depends on your preferred approach to learning. Honey and Mumford (*The Manual of Learning Styles*—see **10.4**) produced a questionnaire to help people identify their 'learning styles'. Their research indicated that we all feel more comfortable with some parts of the learning cycle than with others. The results of their 'Learning Styles Questionnaire' divide responses into four groups, or styles. Each of the styles corresponds to a part of the learning cycle. Look at the style descriptors below. Which ones best describe your preferences?

(a) *Activist*—an activist learner enjoys taking part in role plays and other exercises. On the whole they are ready to give anything a try.

(b) *Reflector*—reflectors are most comfortable in the reflective stage of the cycle. If role play is involved, they much prefer to let the activists get on with it, while they observe and try to make sense of what they are observing.

(c) *Theorist*—theorist learners are much more comfortable with the 'abstract conceptualisation' stage of the cycle. They are interested in cause and effect, and consistency of ideas. They tend to be uncomfortable with ambiguity and want to find explanations for everything.

(d) *Pragmatist*—this is the 'strategic' learner. This learner sees learning anything as a means to an end. They are rarely interested in the finer nuances of principle and theory. They need to know that something works and will help them achieve their aims.

There are two points to remember about learning styles:

(a) They are not fixed. Using a particular approach to learning will make you more comfortable with it.

(b) We all have a hierarchy of preferred styles, so we should not label ourselves, or others, as being only one type of learner.

To summarise, then, experiential learning differs from course-based (or propositional) learning in a number of important ways:

(a) There is no syllabus.

(b) There is no formal assessment.

(c) You decide what is and what isn't important.

(d) You decide what to learn.

(e) You determine the learning process.

(f) You build up a repertoire of beliefs about what will be effective in given situations.

(g) You continue to refine this throughout your professional life.

10.3 How to use your experience to become more proficient

If you are the kind of learner who likes to write things down, then you will probably take to the learning diary approach to experiential learning. This is a very simple and effective way of documenting your learning. Because you have to write things down, it encourages you to deliberate about your experience in more detail than you might otherwise do. If you want to try this approach, divide a piece of paper into four sections horizontally. Label them as follows:

(a) event;

(b) reflection;

(c) conclusions; and

(d) action plan.

You will see that each section corresponds to one of the stages of the learning cycle. When an event takes place that gives you a learning opportunity, write down what you consider to be the main features. In the 'reflection' part, note down your comments on what you did (or didn't do); why things turned out as they did; how you felt about the event. In the 'conclusions' section, write your views on how such a situation should be handled in future. Then try to develop an action plan so that you can see if your new theory of action is effective. Finally, when an opportunity presents itself, try to take it and make a new entry to evaluate your performance and, if necessary, modify it.

Try to keep your entries to no more than one side of A4, otherwise the process becomes too onerous for most people. You should also give a little thought to how you are going to file your entries for ease of reference. Will it be chronologically or by category of event?

If you are not someone who enjoys writing, then you might consider other approaches. For example, you might find a collaborator to discuss your learning with. You may find this a very useful way of getting feedback and support for your learning.

During your training contract, your employer may appoint a mentor, a senior person in the firm who will give you guidance in your early career, or who may provide opportunities for you to try different things. At any rate there is likely to be some kind of structure to your development. The purpose of any of these approaches is to provide you with opportunities that you can learn from.

All learning, but professional learning in particular, is voluntary. Your employer can only supply the framework in which you can learn. Whether you take those opportunities and how effectively you learn from them is up to you.

We hope that you have found the material in this book useful for your course work. We also sincerely hope that you will continue to find it useful when you begin your career as a lawyer. Please regard the skills introduced here as a starting point to your learning, not an end point. Good luck.

10.4 Self-development activities

(1) Think of a learning event which has:

- (a) been a success for you; or
- (b) been a failure for you.

What features can you identify which made for success or failure?

(2) Identify something you need or want to learn. Then identify the activities that will enable you to learn it. When you have carried out any of these activities, write about it using the following framework:
    (a) What happened? What did I do? What did others do? What was the outcome?
    (b) How do I feel the activity went, from my perspective, and why? Focus on what I did, how I felt about it, why I decided to do things the way I did.
    (c) What have I learned about this kind of activity and my ability to perform it effectively?
    (d) What would I do differently in a similar situation?

(3) Work with a colleague and ask them to observe you carrying out a work activity. Then ask them to give you feedback. Log your response to the feedback as in task 2 above. Identify what makes for good feedback.

(4) Agree with a colleague to provide them with feedback on a work activity. Try to apply the 'rules' of giving feedback you identified in task 3 above. Ask them if they have learned or confirmed anything about their performance of the activity from your feedback.

10.5 FURTHER READING

Honey, P., and Mumford, A., *The Manual of Learning Styles* (Maidenhead: Honey, 1986).

Kolb, D., *Experiential Learning: Experience as the Source of Learning and Development* (London: Prentice Hall, 1984).

ANSWERS TO SELF-TEST QUESTIONS

Chapter 2

(1) Describe five techniques for encouraging your client to talk.

Answer:

- Use open questions, especially at the beginning. This encourages the client to use their own words and makes the interaction more like a conversation than an interrogation.
- Invite your client to talk. For example, 'Tell me more about...' 'Please go on...' Again, this is about developing a relaxing conversational style and indicates that you are listening actively.
- Use sympathetic body language. That is, making and maintaining eye-contact, smiling, nodding encouragingly.
- Summarise periodically to check your understanding and to let your client correct any misunderstanding. 'So, the situation so far is...', 'Is that right?' 'Let me just recap so far.'
- Use pauses constructively. Allow your client to think through their answer. Don't be tempted to jump in and put words in their mouth, or close down the question.

(2) What is active listening?

Answer:

- Checking what your client is saying against your framework of understanding. Active listening enables you to modify and refine your conception of the legal position.
- Following up points that you hear with appropriate questioning. Again this enables you to develop your framework of understanding.
- Summarising and reflecting back what you have heard. This is where you pick up and explore cues from your client's language or body language about deeper meanings, hidden agendas, need for the client to get things off their chest, etc, e.g. 'You seem to find this very upsetting....'

(3) Why is non-verbal communication important during an interview?

Answer:

Non-verbal communication refers to posture, gesture, facial expression, pauses, intonation, emphasis. All of these are important because they give valuable clues about how a person feels, their attitude to their situation, to you as their legal advisor and the degree to which they grasp what the issues are.

Don't forget that just as you are reading and interpreting your client's NVC, they will be reading and interpreting yours.

Moreover, most of the information in any kind of face-to-face interaction is non-verbal, so not interpreting it correctly can mean understanding only part of the message.

(4) What does the client need to know at the conclusion of the interview?

Answer:

You need to confirm:

- client's instructions;
- that the client understands what is going to happen;
- the client knows the costs and risks involved;
- the client knows who in the firm to contact;
- what action the client needs to take, if any, and by when it needs to be taken;
- the client knows when they will be contacted and what about;
- the client has no further matters to deal with.

In general, the client must have a clear understanding of what is going to happen and the important issues in their case. They should feel confident about your ability to help them achieve their objectives and confident in your professional competence and conduct.

Chapter 3

(1) What are the key differences between the spoken and written language?

Answer:

The written language enables communication over distance and time. However, it lacks a number of important features of the spoken language:

- *Body language*—a combination of gesture, posture and facial expression indicating the state of mind, level of understanding and attitude of each of the participants;
- *Prosody*—the intonation, emphasis, wordless sounds and pauses which convey important meaning;
- *Immediate response.*

(2) What are the implications of this for effective writing?

Answer:

The implications are that writing has to be carefully planned and checked to overcome the absence of non-verbal communication, ensuring that the intended meaning is understood by the reader as quickly as possible with as little effort as possible.

To achieve this, care must be taken to choose words, construct sentences and develop ideas in paragraphs, which are clear, concise and correct. Punctuation is a poor but necessary way of indicating features of prosody. Learn how to use it properly.

It is important to avoid 'legalese', especially to clients, so you need to be clear about what kind of language is appropriate for each recipient of your written communication. So, you should:

- prefer commonly used (but not colloquial) words over infrequently used ones;
- avoid clumsy and inelegant words and phrases;
- where you can, prefer verb phrases to nominalisations;
- generally use the active verb form rather than the passive;
- omit redundant words and phrases;
- prefer specific words and phrases to vague, general ones;
- avoid using jargon and technical terms unless there is no alternative, and you are sure your reader will understand them;
- vary your sentence length;
- avoid too many subordinate clauses in sentences, they ask too much of your reader's memory;
- use the grammar of standard written English and including correct punctuation;
- spell correctly.

(3) How do you ensure that your writing is appropriate for your audience?

Answer:

This will be determined by a number of factors:

- how well you know the reader will determine how formal you need to be, remembering that you should always maintain a professional approach, even to people you know well;
- how easy they find it to grasp the issues and processes involved;
- their likely attitude to the message. What is their probable response?
- their attitude towards you, the writer;
- the client's understanding of English if it is not their mother tongue;
- what the outcome for them is likely to be.

(4) Outline the structure of a typical paragraph.

Answer:

A typical paragraph contains one topic. It will have a link to the previous paragraph, indicating how the new information connects to the previous ideas. It will have a topic sentence, usually at or near the beginning. This encapsulates the main point of the paragraph. The other sentences amplify, illustrate, challenge or otherwise develop the topic. There is often a link to the following paragraph, usually near the end.

Often, the ideas in a paragraph form a hierarchy of ideas. This is difficult to express in writing, which is linear. Therefore how you plan and construct your paragraphs is critical to understanding your communication.

Chapter 7

(1) What characteristics do all types of negotiation have in common?

Answers:

- Negotiation involves two or more parties who have some common interest which takes them to the negotiating table.
- However, they begin with different objectives, and it is this which has prevented them from achieving an outcome so far.
- Initially, both parties think that negotiation will be a more satisfactory way of resolving their dispute than litigation.
- Both think they can persuade the other to modify their original position.
- Both have some power or influence—real or assumed—over the other's ability to act.
- Most negotiation takes place face to face and is strongly influenced by emotion and attitudes as well as facts and argument.

(2) To be a successful negotiator you need to develop a repertoire of knowledge, skills and attitudes. List these.

Answers:

Knowledge: When preparing your negotiation you need to know the context in which it is taking place:

- Who is the client?
- What relationships are involved?
- How is the litigation funded?
- What is the client risking?
- Who are the negotiators?
- What do you know about your opponent?

You must also know your case inside out – facts and law.

Skills: The key skill is analysis: (facts, law, objectives, interests, the situation as it is unfolding). Also:

- active listening;
- questioning;
- influencing;
- use concessions effectively;
- flexibility;
- creative thinking;
- reflection on your negotiating experience.

Attitudes: Develop an awareness of the negotiation process. This involves:

- building a relationship with the other side;
- moving towards each other;
- use positive language and non-verbal communication;
- acting professionally towards the opponent.

Furthermore, remember it is your client's case, and your negotiation is based on the client's objectives, not yours.

(3) What is meant by the terms 'negotiating style' and 'negotiating strategy'?

Answers:

Negotiating style is interpersonal behaviour—how you communicate with the other party. The negotiation literature identifies two basic styles: competitive and co-operative. Competitive negotiators tend to view negotiation as a game, the aim of which is to score more points than the other side. They come across as tough, dominating, forceful and are unwilling to share interests or move from their opening position.

Co-operatives favour a fair and ethical approach to bargaining. They are willing to share information, seek common ground and move from their opening position. This style is thought to lead to more satisfactory outcomes than the competitive style, which runs the risk of stalemate and deadlock.

The major disadvantage of the co-operative style is that a competitive opponent may interpret it as a sign of weakness and so increase their demands, putting heavy pressure on the co-operative to concede without getting anything in return.

Strategy is the general approach you take to the negotiation. There are two basic strategies—win/lose and win/win. Win/lose is an adversarial strategy. Both parties want as much as they can get from a fixed cake, so that one party's gain is the other's loss. This strategy is thought to underpin both competitive and co-operative styles of bargaining.

Win/win is a problem-solving strategy which sees negotiation as a decision-making process based on underlying common interests and objectives. Parties look to expanding the cake for mutual gain.

(4) What are the four 'syndromes' most likely to derail a legal negotiation?

Answers:

The 'one track' syndrome
The negotiator enters the negotiation with agenda and solution worked out in advance. There is no intention to negotiate, only to impose! Negotiation is a joint decision-making process, and any agenda prepared in advance is only a guide. This syndrome may arise from lack of confidence and the desire to 'control' what is in reality an open-ended, interactive process.

The 'random walk' syndrome
The parties are going round in circles, getting nowhere. No conclusions are reached on any issue, instead the parties jump to a new issue. This may result from insufficient preparation, or the unwillingness of the parties to confront conflict which might arise if a problem area is discussed in depth.

The 'conflict avoidance' syndrome
Parties concentrate on the less important issues rather than those seriously in dispute. Alternatively, a party may be too keen to make concessions. This happens where you want the negotiation to be as painless as possible, or it is important to you that the opponent likes you. The treatment for this syndrome is to remind yourself why you are there. Giving in is not acting in the best interests of your client.

The 'win/lose' syndrome
The parties view the negotiation as an adversarial contest and so use trial tactics, which result in deadlock. The tactics include a refusal to move from an entrenched position,

negative language and body language, and refusal to acknowledge the other side's position. As above, ask yourself why you are there rather than in court.

Chapter 9

(1) What key documents will help you develop your time management skills and what is each used for?

Answer:

- *To-do list*—to show what needs to be done each day and as a way of checking what has and has not been achieved.
- *Time log*—to check how you actually spend your time and how long tasks take to complete.
- *Time management grid*—to help to set priorities by differentiating between important and urgent tasks; to chart what kinds of task you are currently spending your time on.
- *Diary or personal planner*—to note times and dates of appointments, deadlines; to help make sure that preparation for appointments and meetings is done on time; to carry forward to an appropriate time uncompleted tasks; to keep others informed of your activities and whereabouts; to act as a record for billing purposes.

(2) What are the four key elements of project management and how can a Gantt chart help?

Answer:

- Identify all the key elements that make up a project.
- Set appropriate time periods for each activity.
- Show whether an activity is dependent on, or independent of, other activities.
- Link to your personal time management system.

A Gantt chart helps by providing a graphic representation of the key activities of a project. It shows when each element begins and by when it should end. It also shows which elements are dependent on others.

(3) When a task/project is delegated to you what six things is it important to be clear about?

Answer:

It is important to be clear about:

- what the task is;
- what outcomes are being sought;
- what resources are being allocated to the task;
- the time-scales involved;
- who else is involved;
- who is/are the persons accountable for a successful outcome.

(4) What factors does the research of Arnold et al. (1998) show as being the main causes of stress?

Answer:

- *Factors intrinsic to the job*—workload, long working hours, the physical working environment, the effect of your decisions on others, etc.
- *Your role in the organisation*—role ambiguity and conflict, susceptibility to anxiety, etc.
- *Relationships at work*—lack of support from seniors and colleagues, unfriendly atmosphere, too great a level of competitiveness, etc.
- *Career development*—progress not according to expectations, job insecurity, feelings of inequity of treatment, etc.
- *Organisational structure and climate*—degree of influence in the organisation, incompatible climate, etc.

GLOSSARY OF IT TERMS

Application	A computer program which carries out a specific task—e.g., word processor, spreadsheet, database, etc.
Attachment	A file sent with (attached to) an e-mail message; using attachments is an easy way to share documents with other users via the Internet.
Back up	A copy of a file kept in case of damage to your original. It is important to back up your work regularly and save it to a different disk.
Bandwidth	The capacity of a communications connection, most typically that between your computer and the Internet. Think of bandwidth as a water pipe—the bigger the pipe, the more water can flow through it. In the same way, the more bandwidth, the quicker you can receive data from the Internet.
Bitmap	A common form of picture file.
Browser	An application used to browse the World Wide Web such as Internet Explorer and Netscape Navigator.
Bug	Error in a program's coding that causes a malfunction.
Byte	Unit of computer memory. There are one thousand bytes in a kilobyte (kb) and one thousand kilobytes in a megabyte (mb). There are one thousand megabytes in a gigabyte.
Cache	Temporary storage space. Web pages you browse can be stored here to save time downloading them from servers again.
CD-ROM	Compact Disk Read Only Memory—a CD formatted for use by computers for data storage. CD-ROMs are typically 650 mb in capacity, though some can be up to 700 mb.
CD-R	A compact disk that can be written to by a computer equipped with a CD-RW drive. Once written, data is permanently stored on a CD-R, and the disk behaves like a CD-ROM; it cannot be overwritten.
CD-RW	Like a CD-R, but this can be erased and used again. More expensive to buy than CD-Rs and not compatible with as many reader drives.
Clipboard	A temporary storage area used to store text or images that have been 'copied' or 'cut' via an applications Edit menu.
Cookie	A file stored on your computer when you visit some web pages—this can be used to store useful information about you to customise a website to your needs. Some cookies are used for malicious purposes.

Cont.

CPU (Central Processing Unit)	The brain or engine of your computer. Common CPUs are Intel Pentium and AMD Athlon. A processor's capacity is measured in Megahertz (MHz); the higher the MHz, the better.
Crash	A software breakdown which causes data loss.
Desktop	The base of your computer, it provides access to applications, folders and files—on Windows™-based systems the desktop contains icons such as My Computer and Recycle Bin.
Device	Any piece of hardware used by computer. Typical examples are hard disks, CD-ROM drives, printers and scanners.
Dialogue box	A window which pops up to give you information, a warning or to ask a question.
Directory	An area of storage space. In Windows™ environments these are represented by folders.
DOS	Disk Operating System, a basic operating system for the control of files, folders and disks. MS-DOS™ and PC-DOS™ were early DOS-based systems. Contemporary Windows™ systems are based on DOS.
Download	Transfer files from another location to your computer. E.g., obtaining files from university servers or the Internet.
Driver	A file which integrates into the operating system to enable the operating system to control a device such as a printer or video card.
E-business	Similar to e-commerce, but for business-to-business transactions rather than retail.
E-commerce	Using the Internet to carry out transactions, buying, selling and managing finances online.
E-mail	An electronic communication transmitted via the Internet or an internal network.
E-mail address	A person's unique address for e-mail messages e.g., john.smith@internetprovider.com.
Extension	Often invisible in more recent versions of Windows, an extension is the three letters after the full stop in a Microsoft file name. The extension denotes the file type. E.g., letter.doc—the doc denotes MS Word.
File	Any document or set of data stored on the computer, represented by icons.
Firewall	Security system used by companies and individuals to protect their computers and internal networks against unauthorised intrusion from the Internet.
Floppy disk	A diskette which fits into most modern computers, typically with 1.44 mb capacity.
Folder	A filing cabinet on the computer for organising your files.
Freeze	When the computer stops responding to your commands—this can often mean having to switch off and loose unsaved data.
Gif (Graphics Interchange Format)	An image file format often used to show images on the Internet due to the small amount of space taken up by these files.

Cont.

Graphics card	Sometimes called the video card, the graphics card is the interface from your computer's base unit to the screen. The more memory your graphics card has, the better-quality graphics can be displayed.
Hacker	Term given to someone who unlawfully accesses computer systems and files.
Hard disk	Rewriteable storage volume inside your computer—this is the main location for storing files.
HTML (HyperText Markup Language)	A language traditionally used to create web pages.
Icon	A graphical representation of files or programs.
The Internet	The International Network of computers.
Intranet	An organisation's internal network. Recently organisations have selected web-based technology for use on their Intranets, making them look like the Internet.
ISP (Internet Service Provider)	A company providing connections to the Internet and E-mail/Webspace facilities, e.g., Demon Internet, AOL, etc.
JANET	The Joint Academic Network is an Internet backbone for universities and other educational establishments.
Java/Javascript	A programming language often used to enhance the appearance of web pages.
Laptop	Portable computer—sometimes called a notebook.
Load	To transfer a file or program from disk into the computer's RAM so that it can be used.
Memory	Storage space, often synonymous with RAM—Random Access Memory—which is volatile memory used to operate programs and files. The contents of RAM are lost when the computer is switched off; this is why unsaved files are lost.
MODEM	Modulator DeModulator—the device which converts computer data into sound and vice versa, allowing transfer of data and connection to the Internet via telephone lines.
Motherboard	Circuit board inside your computer linking the processor, RAM, disk drives and interface cards together—often thought of as the spinal chord.
Mouse	So named because of its shape and 'tail', the once revolutionary mouse has become an everyday part of the computer interface.
MP3 File (MPEG3)	A multimedia format allowing sound and video to be stored without taking up too much space—commonly used for sharing music via the Internet (often illegally!).
Operating system	Eg., Windows. The operating system is the interface between you or your application programs and your computer's hardware.
Pentium	A brand name for a range of processors manufactured by Intel.

Cont.

Pixel	The computer screen is divided into many little dots called pixels—the appearance of these is altered to form images. If you adjust your computer's resolution, you can choose how many pixels you wish to appear on your screen—the higher resolution, the higher the quality.
Port	An interface (either physical or virtual) for connecting devices or services to.
Printer	Device for getting lots of paper stuck or torn and spilling ink everywhere—some say they can also be used for transferring the contents of the screen to paper!
Processor	See CPU.
Program	See Application.
RAM	See Memory.
Reboot	The process of shutting down and immediately restarting your computer—either by using the computer's Restart command, or by more coercive measures in the event of a freeze.
Recycle bin	When you delete a file—provided it is on your computer and not on the network—it will be placed in the recycle bin, where you can retrieve it if you find you want it after all.
Run	To start a program.
Scanner	Device used for transferring hard copy text and pictures to your computer.
Search engine	An Internet database of web pages that select content based on keywords you specify—e.g., Google.
Server	Any computer which holds files or applications that you access to via a network.
Secure	Secure connections are used for e-commerce and other confidential exchanges over the Internet—a secure connection is often indicated by the presence of a padlock in the bottom of your browser window.
Software	Any application, program or operating system.
Start button	Feature of Windows 95 onwards—gives access to programs, control panel and other features. Curiously one must use the start button to shut down the computer.
Taskbar	Feature of Windows 95 onwards—bar along the bottom of the screen (can be moved) that allows you to click between different programs you are running. In the bottom left of the taskbar is the Start button; in the bottom right is the System Tray which contains the clock and volume control, and links to certain other applications.
Toolbar	Bars of mini-icons used in applications to give quick links to common functions—hovering the mouse over a particular button will normally display a description of its function.
Uninstall	A procedure to remove a program from your computer. Note that this is the only way to remove a program; manually deleting the program's files will cause serious problems to the stability of your computer system.

Cont.

Upgrade	The process of updating a piece of software to a newer version.
URL (Uniform Resource Locator)	Web address http://
Upload	Transfer files to an online location.
Virus	A program that can either simply be annoying or cause serious malicious damage to your computer. Nowadays many virus warnings are sent by e-mail; often there is no actual virus, just a hoax started to waste people's time. Never forward such messages without first checking if they are genuine—one place you can do this is http://www.symantec.com.
Windows	Microsoft's base operating system—so named because it displays your different programs in panels called Windows.
World Wide Web	Shared content in the form of web pages and downloads available over the Internet—you need a browser such as Internet Explorer or Netscape to 'surf' the World Wide Web.

Index

M

N